Kai Fischer

High Performance Light Water Reactor - Next Generation Nuclear Power

Kai Fischer

High Performance Light Water Reactor - Next Generation Nuclear Power

Conceptual Design of a Reactor Pressure Vessel and its Internals for a Supercritical Water-Cooled Reactor

Südwestdeutscher Verlag für Hochschulschriften

Impressum/Imprint (nur für Deutschland/ only for Germany)
Bibliografische Information der Deutschen Nationalbibliothek: Die Deutsche Nationalbibliothek
verzeichnet diese Publikation in der Deutschen Nationalbibliografie; detaillierte bibliografische
Daten sind im Internet über http://dnb.d-nb.de abrufbar.
Alle in diesem Buch genannten Marken und Produktnamen unterliegen warenzeichen-, marken-
oder patentrechtlichem Schutz bzw. sind Warenzeichen oder eingetragene Warenzeichen der
jeweiligen Inhaber. Die Wiedergabe von Marken, Produktnamen, Gebrauchsnamen,
Handelsnamen, Warenbezeichnungen u.s.w. in diesem Werk berechtigt auch ohne besondere
Kennzeichnung nicht zu der Annahme, dass solche Namen im Sinne der Warenzeichen- und
Markenschutzgesetzgebung als frei zu betrachten wären und daher von jedermann benutzt
werden dürften.

Verlag: Südwestdeutscher Verlag für Hochschulschriften Aktiengesellschaft & Co. KG
Dudweiler Landstr. 99, 66123 Saarbrücken, Deutschland
Telefon +49 681 37 20 271-1, Telefax +49 681 37 20 271-0, Email: info@svh-verlag.de
Zugl.: Stuttgart, Universität Stuttgart, Diss., 2008

Herstellung in Deutschland:
Schaltungsdienst Lange o.H.G., Berlin
Books on Demand GmbH, Norderstedt
Reha GmbH, Saarbrücken
Amazon Distribution GmbH, Leipzig
ISBN: 978-3-8381-1130-8

Imprint (only for USA, GB)
Bibliographic information published by the Deutsche Nationalbibliothek: The Deutsche
Nationalbibliothek lists this publication in the Deutsche Nationalbibliografie; detailed
bibliographic data are available in the Internet at http://dnb.d-nb.de.
Any brand names and product names mentioned in this book are subject to trademark, brand
or patent protection and are trademarks or registered trademarks of their respective holders.
The use of brand names, product names, common names, trade names, product descriptions
etc. even without a particular marking in this works is in no way to be construed to mean that
such names may be regarded as unrestricted in respect of trademark and brand protection
legislation and could thus be used by anyone.

Publisher:
Südwestdeutscher Verlag für Hochschulschriften Aktiengesellschaft & Co. KG
Dudweiler Landstr. 99, 66123 Saarbrücken, Germany
Phone +49 681 37 20 271-1, Fax +49 681 37 20 271-0, Email: info@svh-verlag.de

Copyright © 2010 by the author and Südwestdeutscher Verlag für Hochschulschriften
Aktiengesellschaft & Co. KG and licensors
All rights reserved. Saarbrücken 2010

Printed in the U.S.A.
Printed in the U.K. by (see last page)
ISBN: 978-3-8381-1130-8

Contents

Abbreviations   v

Nomenclature   vii

1 Introduction   1
    1.1 Design Concepts for Nuclear Reactors with Supercritical Water   3
    1.2 Aim and Outline of the Study   14

I Fundamental Design Methods   17

2 Mechanical Analysis of the Vessel and its Internals   19
    2.1 Applied Safety Standards   19
    2.2 Load Case Classes   19
    2.3 Design Requirements   21
    2.4 Dimensioning   21
    2.5 Mechanical Behavior   21
        2.5.1 Stress Analysis   21
        2.5.2 Fatigue Analysis   23
        2.5.3 Thermal Strain Analysis   23
    2.6 Thermal Loads   23
        2.6.1 Dimensionless Numbers   23
        2.6.2 Overall Heat Transfer   24
        2.6.3 Heat Transfer for Concentric Annular Gaps   25
        2.6.4 Heat Transfer for Free Convection on Vertical Surfaces   26

3 Flow Analysis of a Backflow Limiter   27
    3.1 Performance Factor   28
    3.2 Pressure Loss Coefficients   28
    3.3 Swirl   28

II Numerical Methods   29

4 Finite Element Simulation   31
    4.1 Theory of Elasticity   32
    4.2 Discretization Method   34
    4.3 Geometry and Grid Generation   37

	4.4	Coupled Thermal-Structural Analysis	39
	4.5	Boundary Conditions	40

5 Computational Fluid Dynamics — 41
- 5.1 Basic Conservation Equations — 41
- 5.2 Turbulence Modeling — 43
  - 5.2.1 Eddy Viscosity Models — 43
  - 5.2.2 $k\text{-}\omega$ SST model — 45
  - 5.2.3 Non-Linear Models — 46
- 5.3 Discretization Method — 46
  - 5.3.1 Interpolation Methods — 47
- 5.4 Numerical Solution Method — 48
- 5.5 Geometry and Grid Generation — 49
- 5.6 Boundary Conditions — 50
- 5.7 Grid Sensitivity Analysis — 52
- 5.8 Evaluation of Turbulence Models — 54

III Concepts and Analyses — 57

6 Reactor Design Overview — 59

7 Design of Core Components — 63
- 7.1 One Pass Core — 63
- 7.2 Modifications for a Two Pass Core — 65
- 7.3 Three Pass Core — 67

8 Internals Design — 75

9 Reactor Pressure Vessel Design — 81

10 Dimensioning of Critical Components — 85
- 10.1 Candidate Materials for the Components — 85
- 10.2 Mechanical Analysis — 87

11 Design Verification Using Finite Elements Methods — 91
- 11.1 Geometry and Numerical Model — 91
- 11.2 Boundary Conditions — 94
- 11.3 Temperature Distribution and Deformation Analysis — 96
- 11.4 Evaluation of Stress Intensities — 98

12 Fluidic Optimization of the Backflow Limiter — 103
- 12.1 Design of the Backflow Limiter — 106
- 12.2 Optimization Procedure — 107
- 12.3 Numerical Model in STAR-CD — 108
- 12.4 Results of the Optimization — 110
- 12.5 Characteristic of the Backflow Limiter — 115

13 Conclusions	119
Bibliography	123

Appendix 131

A KTA Guidelines 133
- A.1 Design Fatigue Curves according to KTA 3201.2 134
  - A.1.1 Ferritic Steels 134
  - A.1.2 Austenitic Steels 135

B RPV Assembly for the HPLWR 137
- B.1 Notation 137
- B.2 RPV Design for the One Pass Core 138
- B.3 RPV Design for the Two Pass Core 139
- B.4 RPV for the Three Pass Core 140

List of Abbreviations

AB	Anormale Betriebsfälle = Anomalous Operational Load Cases
ADS	Automatic Depressurization System
AECL	Atomic Energy of Canada Limited
AF	Auslegungsfälle = Design Load Cases
AGR	Advanced Gas-cooled Reactor
AMG	Algebraic MultiGrid
ANL	Argonne National Laboratory
ANSYS	ANalysis SYStem
APWR	Advanced Pressurized Water Reactor
ASME	American Society of Mechanical Engineers
BWR	Boiling Water Reactor
CAD	Computer-Aided Design
CANDU	CANada Deuterium Uranium
CATHARE	Code Avancé de Thermohydraulique pour Accidents de Réacteur à Eau (nuclear safety analysis code for PWR)
CATIA	Computer Aided Three dimensional Interactive Application
CFD	Computational Fluid Dynamics
CG	Conjugate Gradient
CRGA	Control Rod Guide Assembly
DIN	Deutsche Industrie Norm
DOF	Degrees of Freedom
EPR	European Pressurized Water Reactor
ERCOFTAC	European Research Community On Flow, Turbulence, And Combustion
EURATOM	EURopean ATOMic Energy Community
FA	Fuel Assembly
FEM	Finite Elements Method
FFPP	Fossil Fired Power Plants
GFR	Gas-Cooled Fast Reactor
GIV	Generation IV International Forum
HDR	HeissDampfReaktor
HPLWR	High Performance Light Water Reactor
IKET	Institute for Nuclear and Energy Technologies
INL	Idaho National Laboratory
KTA	Safety Standards of the Nuclear Safety Standards Commission in Germany
KWU	KraftWerksUnion
LFR	Lead-Cooled Fast Reactor
LOCA	Loss of Coolant Accident
LWR	Light Water Reactor

MARNET	Marine and Offshore
MSR	Molten Salt Reactor
NASA	National Aeronautics and Space Administration
NB	Normale Betriebsfälle = Normal Operational Load Cases
NF	Notfälle = Emergencies
PF	Prüffälle = Test Load Cases
PWR	Pressurized Water Reactor
PWR-SC	Pressurized Water Reactor-Supercritical Conditions; SDWR (german)
QUICK	Quadratic Upwind Interpolation of Convective Kinematics
RPV	Reactor Pressure Vessel
RSM	Reynolds Stress transport Models
SC-PWR	Pressurized Water Reactor, cooled and moderated by SuperCritical water
SCLWR	SuperCritical Light Water Reactor
SCLWR-H	High temperature SuperCritical Light Water Reactor
SCOTT-R	SuperCritical Once-Through Tube Reactor
SCR	Supercritical steam Cooled Reactor
SCWR	Supercritical Water-Cooled Reactor
SF	Schadensfälle = Accidents
SFR	Sodium-Cooled Fast Reactor
SIMPLE	Semi-Implicit Method for Pressure-Linked Equations
SST	Shear Stress Transport
STAR-CD	Simulation of Turbulence in Arbitrary Regions
SWR-1000	SiedeWasserReaktor (1000 MW_e)
UD	Upwind Differencing
VHTR	Very-High Temperature Reactor
WWER	Water-cooled, Water-moderated Energy Reactor

Nomenclature

Latin symbols

a	m²s⁻¹	Temperature diffusivity
A	m²	Transfer area
$A_{element}$	m²	Area of the triangle plane
A_{wall}	m²	Mean transfer area for the tube wall
d	m	Displacement
d_h	m	Hydraulic diameter
d_i	m	Inner pipe diameter
d_o	m	Outer pipe diameter
D	-	Cumulative damage factor
D_i	m	Inner diameter of the backflow limiter section
D_o	m	Outer diameter of the backflow limiter section
E	Nmm⁻²	Young's modulus
F	Nmm⁻²	Peak stress
F	m²	Flow cross section area
F_a	N	External vector work force
F_k	N	Vector of the nodal forces
g	ms⁻²	Earth gravity
G	Nmm⁻²	Shear modulus
h	m	Element thickness
k	m²s⁻²	Turbulent kinetic energy
k	Wm⁻²K⁻¹	Heat transition coefficient
K	-	Resistance coefficient (Eqn. 3.2)
l	m	Thermal expansion of the material
L	m	Characteristic length (hydraulic diameter)
l_b	m	Characteristic length (volume height)
l_{tube}	m	Pipe length
l_0	m	Initial length at 0 °C
n	-	Number of cycles to failure
$\hat{n}$	-	Allowable number of cycles
p	Pa, bar	Pressure
p_D	MPa	Design pressure
P	MW	Power
P_b	Nmm⁻²	Primary bending stress
P_k	-	Rate of Production of turbulent kinetic energy
P_m	Nmm⁻²	General primary membrane stress
Q	Nmm⁻²	Thermal secondary stress
q	Wm⁻²	Heat flux between the fluid and the surface
r	m	Radius in Φ-direction

R_{mRT}	Nmm^{-2}	Minimum tensile stress at room temperature
R_{mT}	Nmm^{-2}	Minimum tensile stress at elevated temperature
$R_{p0.2T}$	Nmm^{-2}	0.2 % elevated temperature proof stress
S	kgm^2s^{-1}	Swirl
S_a	Nmm^{-2}	One-half the value of cyclic stress range
S_m	Nmm^{-2}	Allowable material stress intensity
t	°C	Given temperature of the component
t_0	°C	Temperature of 0 °C
T_D	°C	Design temperature
U	m	Perimeter
u	ms^{-1}	Velocity in x direction
$\bar{u}$	ms^{-1}	Time averaged mean velocity in x direction
u'	ms^{-1}	Velocity fluctuation in x direction
v	ms^{-1}	Velocity in y direction
$\bar{v}$	ms^{-1}	Mean fluid velocity (Eqn. 3.2)
$v_{peripheral}$	ms^{-1}	Peripheral velocity
v_s	ms^{-1}	Mean fluid velocity (Eqn. 2.7)
V	m^3	Volume
V_k	m	Displacement vector
w	ms^{-1}	Velocity in z direction
w	-	Gaussian filter function
X	N	Occurring body forces in x direction
y^+	-	Dimensionless wall distance
Y	N	Occurring body forces in y direction
Z	N	Occurring body forces in z direction

Greek symbols

α	$Wm^{-2}K^{-1}$	Heat transfer coefficient
α_T	K^{-1}	Thermal expansion coefficient
β	K^{-1}	Volumetric thermal expansion coefficient
Δ	-	Difference
δ	-	Virtual displacement
δ_{ij}	-	Kronecker delta
δ_{wall}	m	Wall thickness
γ	-	Shear rate
λ	$Wm^{-1}K^{-1}$	Thermal conductivity of the material/fluid
μ	-	Poisson number
μ_t	-	Turbulent viscosity
∇	-	Nabla operator
ν	m^2s^{-1}	Kinematic viscosity
ω	ms^{-1}	Turbulent frequency
ρ	kgm^{-3}	Density
σ_{al}	Nmm^{-2}	Allowable stress intensity
σ_{max}	Nmm^{-2}	Maximum observed stress
$\sigma_{V, v. Mises}$	Nmm^{-2}	Equivalent normal stress, based on the theory of von Mises
σ_x	Nmm^{-2}	Normal stress in x-direction

σ_y	Nmm^{-2}	Normal stress in y-direction
σ_z	Nmm^{-2}	Normal stress in z-direction
Σ	-	Performance factor (Eqn. 3.1)
τ_{xy}	Nmm^{-2}	Shear stress in xy-direction
τ_{xz}	Nmm^{-2}	Shear stress in xz-direction
τ_{yz}	Nmm^{-2}	Shear stress in yz-direction
ε	%	Strain
ε	m^{3}s^{-3}	Dissipation rate
ξ	-	Pressure loss coefficient (Eqn. 2.15)
ζ	-	Pressure loss coefficient (Eqn. 3.3)

Indices

A	Reverse flow direction
B	Regular flow condition
el	Electric
i	Section i
m, T	Average temperature
t	Transpose of the matrix
th	Thermal
w	Wall
x	x direction
xy	Planar surface in x and y direction
xz	Planar surface in x and z direction
y	y direction
yz	Planar surface in y and z direction
z	z direction

Dimensionless numbers

Gr	Eqn. 2.9	Grashof number in the case of natural convective heat transfer
Nu	Eqn. 2.8	Nusselt number
Pr	Eqn. 2.6	Prandtl number
Pr$_t$	-	Turbulent Prandtl number
Ra	Eqn. 2.10	Rayleigh number
Re	Eqn. 2.7	Reynolds number

1 Introduction

In view of continuous growth of the human population with industrial expansion in developed countries and the need to accelerate the industrialization of developing countries, there is no doubt that there will be an increasing demand for energy in the next decades (see Figure 1.1, photograph made from several views of the whole world, showing city lights and areas of greater population). The deregulation of electric utilities, environmental effects and the desire to reduce global warming have an additional impact on the future global energy mix. Nuclear power is seen as a viable option to satisfy this increasing demand for generating safe, clean and economic electricity.

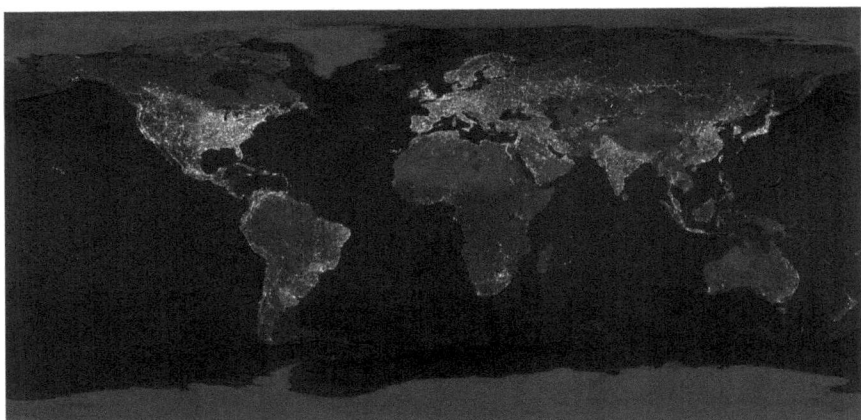

Figure 1.1: NASA photograph made from several views of the whole world, showing city lights and areas of greater population (Nov. 2007).

Evidence of this trend is already seen in the United States, Finland, France, England, South Korea, China and Japan where several new nuclear power plants are planned or under construction. To fortify this trend of increased use of nuclear power for the future, an international co-operation in research for a future generation of nuclear energy systems has been established. In 2000, nine countries (Argentina, Brazil, France, Japan, the Republic of Korea, the Republic of South Africa, Switzerland, the United Kingdom and the United States) agreed in a joint research and development framework in nuclear power, called Generation IV International Forum (GIV, [90]). In 2003, the European Atomic Energy Community (EURATOM) joined this International Forum. The objective is to support and develop innovative systems providing enhanced safety, minimal waste, proliferation resistant and highly economical nuclear energy systems within a time frame of 15 to 20 years. Six different reactor concepts have been chosen to be investigated until 2030. Among them quite known types like the lead-cooled fast reactor (LFR), and the sodium-cooled fast reactor (SFR) but also gas-cooled types like the gas-cooled fast reactor (GFR), and the very-high temperature reactor (VHTR), and the more exotic molten salt reactor (MSR).

The sixth concept involves a further improvement in the economics and efficiency of the well established Light Water Reactor (LWR) similar to the improvements made in fossil fired power plants. This type is called supercritical water-cooled reactor (SCWR) and involves a water-cooled high temperature and high pressure LWR working above the thermodynamically critical point of water (see Figure 1.2). In Europe, investigations on the SCWR have been integrated into a joint research project which is called High Performance Light Water Reactor (HPLWR) and which is co-funded by the European Commission.

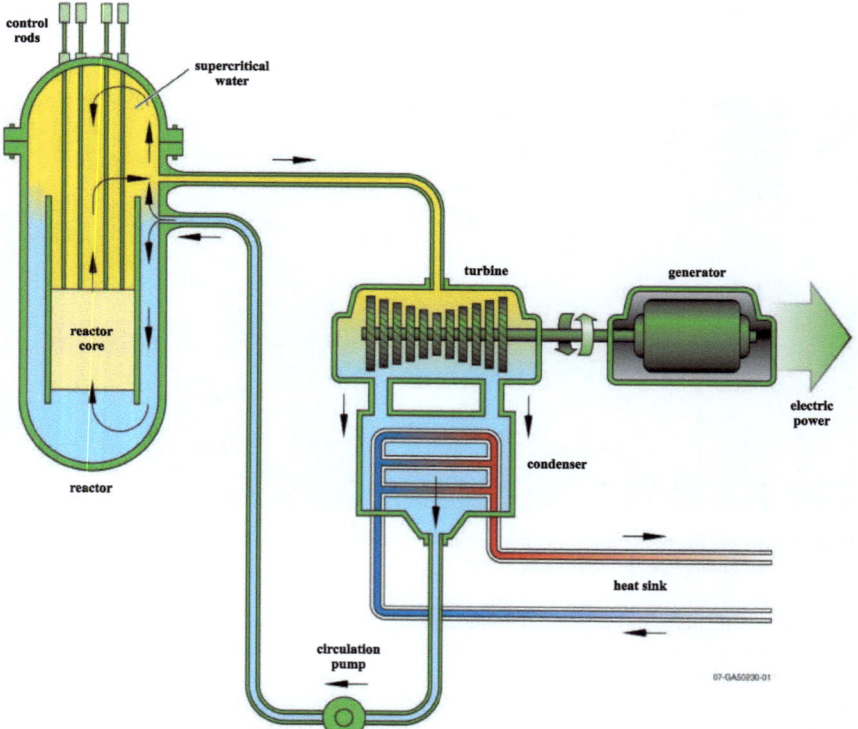

Figure 1.2: Scheme of the supercritical water-cooled reactor with its once-through direct cycle according to [90].

A consortium of European research institutes and industrial partners together with the University of Tokyo has assessed the major issues for this new reactor concept. The main objective of this project is the evaluation of the potential and merit of the HPLWR to help sustain the nuclear option, with a focus on competitively priced electricity and an efficiency of around 44 %. Because of size reduction of components and buildings compared to current LWRs, the reactor concept shall have low construction costs in the vicinity of 1000 Euros per kW$_e$, according to Bittermann et al. [7]. Additionally, low electricity production costs are expected with a targeted 3 to 4 cents per kWh.

The plant characteristics have been defined by Squarer et al. in 2003 [83] within the HPLWR Phase 1 project in 2000 to 2002. The system includes a supercritical coolant pressure of around 25 MPa and a coolant heat up from 280 °C feedwater inlet to more than 500 °C for the steam outlet. Water enters the reactor as liquid and exits as single-phase high-pressure steam. The turbines can be driven directly by the heated coolant making steam-water separation obsolete. In 2006, a second phase called HPLWR Phase 2 was initiated focusing on the assessment of the critical scientific issues and the technical feasibility. It involves the design of a feasible core with a heat up of more than 200 °C for both thermal and a fast neutron spectrum. Those boundary conditions are also expected to have a major impact on the design of the reactor pressure vessel (RPV) and the internals due to the higher pressure and temperature compared to conventional LWR designs. A feasible reactor pressure vessel and suitable internals shall be designed. Special interest has to be contributed to the cooling of the core under all operational and accidental conditions, i.e. a loss of coolant accident (LOCA) for the cold leg of the RPV. Preferably, passive safety systems and components have to be developed to fulfill this task. Other points include material corrosion behavior, investigation of heat transfer deterioration under supercritical conditions and a possible plant layout.

Within the scope of this thesis, feasible designs for the RPV and the internals for such an HPLWR for three different core concepts are presented and evaluated. Additionally, a backflow limiter is designed and analyzed serving as a necessary passive safety device for the inlets of the RPV in the case of a LOCA. The work has been performed at the Institute for Nuclear and Energy Technologies (IKET) at Forschungszentrum Karlsruhe GmbH within the HPLWR Phase 2 project working packages 1 and 2, which include design and integration and core design, respectively (Schulenberg et al. [77]). The design studies have been reviewed by the consortium partner AREVA NP and will contribute to a thorough assessment of the HPLWR concept in order to determine its future potential.

1.1 Design Concepts for Nuclear Reactors with Supercritical Water

With the development of electricity generation by nuclear power, a variety of reactor concepts were investigated. Among them LWR such as the pressurized water reactor (PWR), and the boiling water reactor (BWR), but also supercritical-pressure water cooled reactors (SCWR). The first two types work with live steam at 7.8 MPa and boiling temperatures of up to 286 °C in a direct or indirect cycle, yielding efficiencies over 30 %. These two LWR designs dominate today's nuclear electricity production with over 360 units built and several under construction. The newest design, the European Pressurized Water Reactor (EPR, [31]) features a thermal power of 4250 MW$_{th}$ and a net efficiency of 36 %. Compared to other large scale power plants this efficiency, however, is quite low with almost no improvements since the 1960's. Fossil fired power plants (FFPP), on the other hand, have increased their efficiencies significantly in the same period using superheated steam with increased live steam temperatures and pressures up to 600 °C and 30 MPa, respectively. Today, this stepwise increase yields more than 46 % net efficiency for new coal fired power plants using supercritical steam conditions.

Starting in the 1960's, the application of such highly efficient steam cycle technologies has been seen as a huge potential for further improvements of light water reactors and led to the first SCWR design concepts. As in a BWR, the steam at supercritical conditions is fed directly to the high pressure turbine, with no closed primary cycle of a PWR necessary. The higher steam enthalpy at the turbine inlet allows the reduction of the steam mass flow rate and therefore the size of the turbine and the overall steam cycle with its different components. Together with the advantage of the increased efficiency, this approach simplifies the overall plant design and therefore reduces the electricity production costs. Additionally, the coolant water changes its phase continuously from liquid to steam without boiling at those conditions, excluding the occurrence of a boiling crisis in the nuclear reactor core, this being an important safety advantage compared to the BWR design.

The physical properties of supercritical water also include some challenges which do not occur for subcritical conditions. The coolant water is normally also used for moderation purposes to maintain a thermal neutron spectrum, where a higher density of the water results in better moderation. Density differences of the coolant[1] throughout the core will exceed those of the BWR, due to the high steam temperatures. This results in a lack of moderator, which has to be compensated for. The much higher enthalpy rise in the SCWR compared to BWR and PWR designs also implies the use of high temperature materials and special core concepts in order to avoid hot-spots and structural damages in the core. The higher temperatures and operation pressure also have a major impact on the design of the RPV and the internals. The coolant flow rate of the reactor is quite small due to the high enthalpy rise and the once-through cycle. A postulated break of one of the inlet feedwater lines could cause a flow reversal in the core, despite opening the depressurization valves of the steam lines resulting in an overheating core. Appropriate safety measurements have to be foreseen to maintain cooling of the core for such transients.

Nuclear reactors with superheated steam were evaluated during the 1950's and 1960's by Westinghouse (1957-1966), General Electric (1959), and AEG-Telefunken (1961-69) when today's LWR design was not yet established. These older concepts were found to be principally feasible, although not necessarily economically competitive. Main differences between the concepts involved the choice of an appropriate steam cycle and the type of moderator. The operation at high pressure and high cladding temperature together with heat-up of the coolant water above the critical point inside the reactor was a concern. It was thought that the rapid change of the physical properties near the critical point (Wagner and Kruse [94]) promoted instabilities in the flow, heat transfer, and reactivity. The possible deposition of radioactivity in the external systems of a direct cycle led to several concepts with indirect cycles. They had the disadvantage of more complex reactor systems which made them less economically competitive.

The concepts studied by Westinghouse involved a light water moderated and supercritical steam cooled reactor with direct and indirect cycle named SCR (1957, Oka [64]), a direct cycle pressure tube design with graphite moderator which was called SCOTT-R

---

[1] The occurring density changes are of a factor of over 8 compared to a factor of 3 for BWR designs.

1.1 Design Concepts for Nuclear Reactors with Supercritical Water

for supercritical once-through tube reactor in 1963 [88] and a modified PWR with a two-loop indirect cycle, which was cooled and moderated by supercritical water (SC-PWR, 1966, Wright and Patterson [97]). General Electric used heavy water as moderator for its once through, light water cooled reactor described in the US Atomic Energy Commission report from Hanford Laboratories (1959, [33]). A review by Argonne National Laboratory (ANL) for the different concepts favored a direct-cycle approach, since it offered the highest possibility for economic power, but pointed out the unresolved problems of deposition of radioactivity in the external systems and material issues due to the hostile environment (Marchaterre and Petrick [56]). The AEG-Telefunken reactor concept studied and projected a superheated BWR termed HDR[2] (Traube and Seyfferth [89]), which worked in subcritical regime with an operation pressure of up to 9 MPa. The core consisted of two different types of fuel assemblies; evaporator fuel assemblies with a steam separator above the core and following superheater assemblies for the saturated steam with an outlet temperature of 430 °C. The coolant passed four times through the core before it left the reactor. As a summary, it can be stated that the presented concepts featured steam outlet temperatures as high as 621 °C with net efficiencies ranging from 30 to 43.5 %, but were disregarded and abandoned in the 1970's to pursue the further developed and commercially more promising BWR and PWR.

New supercritical-pressure reactor concepts for the innovation of LWR emerged in the 1990's from Russia, Canada and Japan. They are also enforced within the Generation IV International Forum, which was initiated in 2000 [90]. The SCWR has been chosen as one of six nuclear reactor technologies with high potential for long-term deployment between 2020 and 2030. The partners investigate different concepts which have to fulfill the agreed conditions.

The Russian concept relies on an integral type PWR called B-500SKDI [82] which is similar to the one abandoned by Westinghouse in the 1960's, but originates from the Russian counterpart to the PWR, the VVER[3]. Differences include the cooling of the core by natural circulation and the implementation of the steam generators inside the reactor pressure vessel. The coolant outlet temperature is quite low with approximately 380 °C creating a thermal power of 1350 MW_{th} and resulting in a gross thermal efficiency of 38.1 %. An advantage of this concept is the ommitance of main circulation pumps, primary pipings, accumulators, and outside steam generators.

The Canadian AECL[4] design concept CANDU-X (Bushby [13], Khartabil et al. [47]) is based on the successful CANDU[5] reactor, which features a heavy water moderated pressure tube design. Several concepts are studied, ranging from an extension of the present CANDU reactor with an indirect cycle and natural convection circulation of the primary coolant to a dual cycle where supercritical water exits the core and feeds directly into a very high pressure turbine combined with a steam generator for the exhaust steam of the turbine. Core outlet temperatures range from 400 to 625 °C depending on the concept, giving a thermal power of 930 to 2536 MW_{th}. Research includes applicable safety systems, supercritical water thermal-hydraulics, and materials compatibility.

In the 1990's the University of Tokyo in Japan started the development of a design

---

[2]HeissDampfReaktor
[3]Russian abbreviation for Vodo-Vodyanoi Energetichesky Reactor or WWER, which can be translated as water-cooled, water-moderated energy reactor and is similar to a pressurized water reactor design
[4]Atomic Energy of Canada Limited:Manufacturer of the CANDU nuclear reactor.
[5]Pressurized heavy water reactor: CANada Deuterium Uranium.

concept for a light water moderated and cooled once-through cycle reactor operating at supercritical pressure [65], [66]. Both, a thermal reactor called supercritical light water reactor (SCLWR) with a coolant outlet temperature of around 400 °C and a high temperature version (SCLWR-H) featuring a coolant temperature of 508 °C are investigated. The core consists of open lattice fuel assemblies in an hexagonal arrangement. Water rods inside the fuel assemblies supply the moderator water. The RPV and control rods are similar to current PWR designs, the vessel wall is cooled by inlet coolant (280 °C) as for a PWR pressure vessel to prevent contact with the hot coolant. The design of the containment and the necessary safety features are adopted from the BWR. Due to the physical properties of supercritical water, it is not possible to measure the water level inside the RPV, as for BWR. Oka et al. [66] suggest to measure the inlet flow rate for the core instead.

A similar approach has been chosen in the United States, where the GIF SCWR program has been led by INEEL[6] and incorporates a reference design which focuses on a large-size, direct-cycle, light water cooled and moderated, base load operation plant (Leading author: P.E. MacDonald [55]). For the operating pressure and core outlet temperatures of 25 MPa and 500 °C, respectively, fuel assemblies with large square water rods with downward flow are used to provide adequate moderation in the core. This arrangement provides a thermal power of 3575 MW_{th} resulting in the recommended thermal efficiency of 44 %. Additional work includes the design of the containment and safety systems.

High Performance Light Water Reactor

In Europe, investigations on a supercritical light water reactor are integrated into a joint research project called HPLWR with ten partners from eight European countries. A first investigation has been carried out within the 5th Framework Program to assess the concept and plant characteristics. They included a supercritical coolant pressure of around 25 MPa and a coolant heat-up from 280 to 500 °C (Squarer et al. [83]). The high temperature steam is directly fed to the high pressure turbine in a once-through cycle (Bittermann et al. [9]).

A first fuel assembly design for a HPLWR with a thermal neutron spectrum has been presented by Dobashi et al. [20] and featured hexagonally arranged fuel pins in an assembly box, cooled by rising coolant and equipped with additional water rods to flatten the axial power profile. The moderator water in those rods flowed downwards from the top of the core in a counter-current arrangement to the rising coolant in the assembly boxes. Control rods were inserted from the top into those water tubes. The radial power distribution of this arrangement was rather non-uniform, requiring different enrichments for the fuel rods in each assembly to homogenize the heat-up.

Several core design studies involving square and hexagonal fuel pin arrangements with additional water tubes and different moderator rods were performed (Yamaji et al. [100], Cheng et al. [17], Joo et al. [44]) to homogenize the power distribution and to optimize the coolant heat-up. Hofmeister et al. [41] performed a design study using those different assemblies. They found that a square arrangement with 40 fuel pins and a single moderator box in the center of the assembly, similar to BWR assemblies, had the highest

---
[6]Idaho National Laboratory

1.1 Design Concepts for Nuclear Reactors with Supercritical Water

power density with the lowest possible fuel enrichment. The design featured cross shaped control rods, which were inserted from the top into the water tubes. Nine of these small assemblies were combined into a common 3 × 3 assembly cluster with dimensions similar to the ones applied in PWRs.

A cross section of this assembly cluster with moderator boxes and control rods can be seen in Figure 1.3 on the right side. On the left side of Figure 1.3 an assembly cluster with a common head and foot piece for easier handling during revisions is shown. Moderator water flows downwards through gaps between assembly boxes and through the moderator boxes. A steam plenum above the core collects the hot coolant rising inside those fuel assemblies and supplies it directly to the turbine. One major challenge of all SCWR core design concepts is the higher enthalpy rise compared to designs with subcritical water. For the HPLWR, the enthalpy rise from inlet feedwater to outlet live steam is almost 2000 kJkg^{-1}, exceeding subcritical water operated reactors by more than a factor of 10. The hottest sub-channel of the core, which is the relevant design criterium for the material limit, can be much higher than the average temperature resulting from the nominal enthalpy rise.

This deviation can be expressed by hot channel factors, which include for example the non-uniform power profile in the core, uncertainties in fuel composition and distribution, water density distribution, neutron leakage or control rod positioning. Uncer-

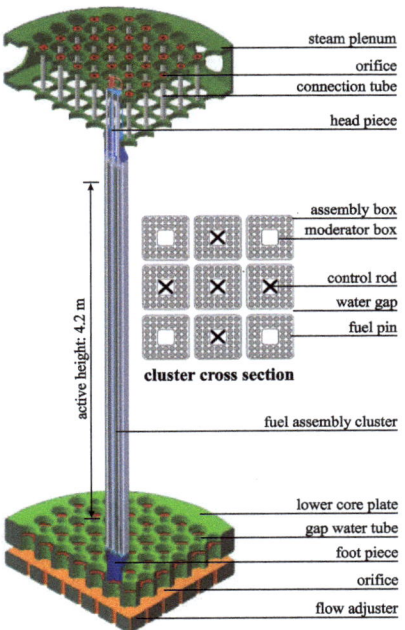

Figure 1.3: Quarter section of the core with one HPLWR fuel assembly cluster [41].

tainties and allowances for operation will also cause hot spots and have an impact on the hottest channel. Schulenberg et al. [78] estimate an overall hot channel factor of approximately 2 to be multiplied with the average enthalpy rise to yield the local maximum enthalpy rise. In 2007, Schulenberg et al. [76] presented several core design options to overcome this issue for the HPLWR. Those designs can be classified by their coolant flow path in the core and consequently are referred to as single, two and three pass core concept, indicating the change of flow direction during heat-up of the coolant. Figure 1.4 illustrates the coolant flow path inside the RPV for all three concepts, as shown in [76].

Starting from a conventional PWR design with an increased pressure and core exit temperature of 25 MPa and 380 °C, the single pass core concept resembles the PWR flow path with feedwater supply at the bottom of the core. The heated coolant from all fuel assemblies is collected at the top of the core and then exits the pressure vessel. Due to the chosen average exit temperature slightly below the pseudo-critical temperature of 384 °C (at 25 MPa), it is possible to run a sub-channel of the coolant at a significantly higher exit enthalpy, while reaching only slightly higher exit temperatures. This phenomenon is contributed to the pronounced peak of the specific heat of water at a temperature of 384 °C. Vogt et al. [93] presented in 2006 such a core design as a near term application of supercritical water technologies, called PWR-SC. Using the fuel assembly design by Hofmeister et al. [41] with 88 clusters, the coolant mass flow is heated up from 280 to 380 °C resulting in a thermal power of 2000 MW_{th}. Each cluster has been equipped with an inlet orifice to adjust the individual coolant mass flow for a homogeneous temperature distribution at the core outlet. With this method, the maximum core exit temperature for the hottest subchannel reaches only 416 °C. Moderator water is flowing downwards in the water boxes inside the fuel assemblies and in gaps between the assemblies. Advantages of this design compared to current PWR include 2 % higher net efficiency, size reduction of the primary loop and less auxilary power for the primary pumps. One drawback is the use of an indirect cycle, due to the liquid condition of the supercritical water at the exit, so that a second loop with steam generators and superheaters is neccessary.

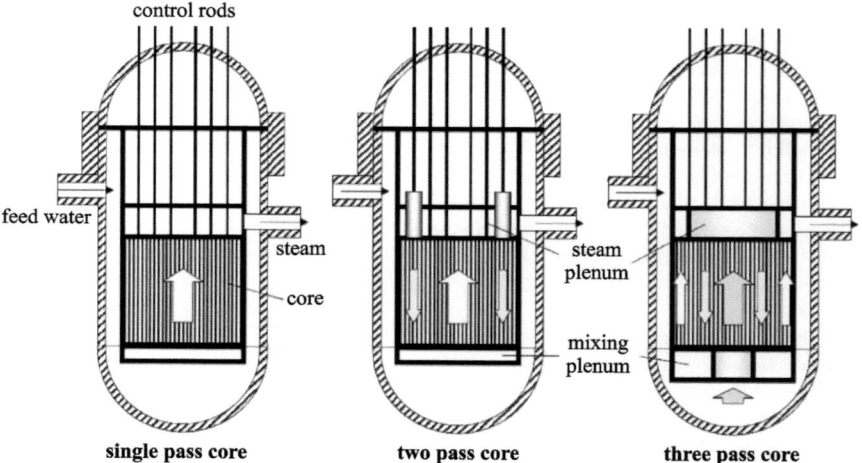

Figure 1.4: Core design concepts for the HPLWR with multiple heat up steps [76].

For a direct cycle, the core exit temperature has to be increased beyond the pseudo-critical temperature, so that the generated steam can be fed directly to a high pressure turbine. The coolant must now be heated up in steps with intermediate mixing to keep the local hot channel temperature within the material limit. The simplest approach is a two-step heat-up (see Figure 1.4), which has been proposed by Kamei et al. [45] and Yamaji et al. [99] in 2005. The design was called Super LWR, and featured a once-through direct cycle without water-steam separators and recirculation pumps. The core consists of 121 square fuel assemblies, each with 300 fuel rods and 36 square water rods inside the fuel rod array. An additional 24 rectangular water rods surrounds the fuel assembly. In the outer section of the core, the downwards flowing coolant is preheated in 48 fuel assemblies to obtain around 380 °C in the lower plenum. To homogenize the temperature, it is mixed with the moderator water, which flows downwards in the

1.1 Design Concepts for Nuclear Reactors with Supercritical Water

core using special moderator rods. From the lower plenum, the coolant rises in 73 fuel assemblies in the inner section of the core representing a superheater to exit with an average temperature of 500 °C, yielding a total thermal power of 2744 MW$_{th}$. A peak cladding surface temperature of 732 °C has been predicted for a further optimized core reaching 530 °C core exit temperature by Yamaji et al. [100] not including allowances and uncertainties for operation. This local peak temperature already exceeds the limit of available cladding materials. Schulenberg et al. [74] suggest to reduce the core average exit temperature to around 430 °C to match the creep and corrosion limits of stainless steel.

To reach the high net efficiency of 44 %, recommended in the GIF program, core exit temperatures have to be around 500 °C. Comparable coolant heat-up with a similar enthalpy rise can be found for supercritical fossil fired power plants. Here, coolant heat-up is done in three steps with one evaporator and two superheaters. For the application in a reactor core, the coolant has to be heated in an evaporator with upward flow, mixed in a steam plenum above the core, followed by a second heat-up in a superheater with downward flow, and a third step with upward flow again after mixing in a second mixing plenum below the core.

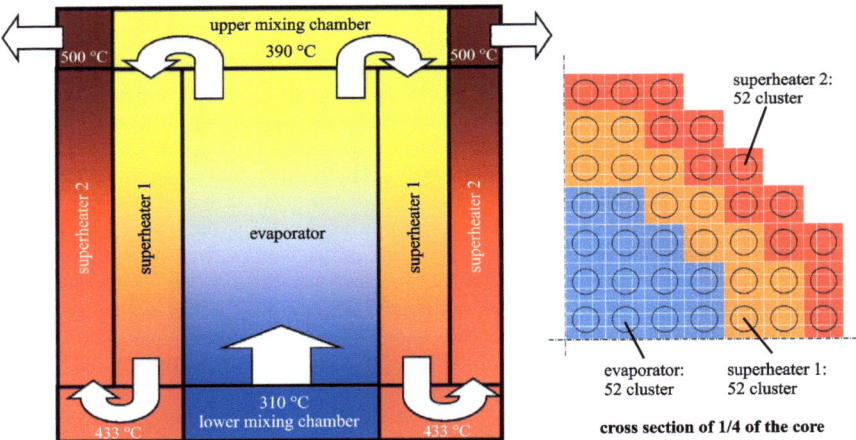

Figure 1.5: Three pass core concept with predicted temperatures for each heat-up step (left), and cross section of the core with the conceptual arrangement of the fuel assemblies (right), according to Schulenberg et al. [78].

In 2006, a possible core design for a HPLWR has been proposed by Schulenberg et al. [78] using a total of 156 fuel assemblies based on the design by Hofmeister et al. [41] and an overall coolant mass flow of 1160 kgs^{-1} to obtain a thermal power of 2188 MW$_{th}$. A first heat-up is provided by heat exchange from fuel assemblies to the downwards flowing moderator water in gaps between the assemblies and in moderator boxes (see scheme in Figure 1.5 on the left side). The moderator water mixes in the lower mixing chamber with feedwater from the downcomer to an average inlet temperature of 310 °C before entering the center region of the core. The coolant is heated up from 310 to 390 °C by 52 clusters in the evaporator before being mixed by the coolant jets in the inner part of the upper mixing chamber. Another 52 clusters surround the evaporator clusters from the superheater 1 with downward flow, heating the coolant to 433 °C. After a second mixing in the outer lower mixing plenum, the rising coolant is heated in superheater 2 to 500 °C by another 52 fuel assemblies arranged in the periphery of the core. A cross section of the core with the conceptual arrangement of the fuel assemblies can be seen in Figure 1.5 to the left. With a proposed peak cladding temperature of 620 °C, this design would stay within material limits of available cladding materials.

Reactor Pressure Vessel and Internals Design

One of the first conceptual designs for the vessel published by Kataoka et al. [46] in 2003 referred to an output of 950 MW_e and featured dimensions similar to a PWR. A preliminary dimensioning of the shell suggested an inner diameter of about 4.3 m, a total height of 15 m and a wall thickness of 0.39 m. The inside of the RPV wall is cooled by feedwater from the inlet, while the hot plenum is covered with a thermal insulator to keep the supercritical steam separated from the surrounding coolant. A disadvantage of this design is the lack of insulation of the hot coolant pipe connection from the hot box to the outlet and the RPV inner wall. In this case, thermal stresses are likely to occur, leading to a possible malfunction of the closure-head seal.

Buongiorno [11] and Buongiorno and MacDonald [12] used a similar approach for their RPV design, but for a higher electrical output of 1600 MW_e. The shell dimensioning has been performed using the ASME Code rules; its structural performance was validated using three-dimensional finite-element analyses. The results yield a height of 12.40 m, an inner diameter of 5.30 m and a maximum wall thickness of 0.46 m for the vessel. In addition to the RPV wall, the hot coolant pipe connection is cooled with feedwater, preventing contact between the hot pipes and the outlet nozzles. In both concepts, two inlets and two outlets are foreseen. Buongiorno and MacDonald [12] also presented a design for the vessel internals using conventional components of a PWR design such as lower core support plate, core former, core barrel, upper core support plate, and upper guide support plate with control rod guide tubes. Calandria tubes at the top of the core supply the downwards flowing moderator water for the water rods. Preliminary stress analyses performed by Westinghouse [55] showed that the high temperature drop across the wall caused unacceptable thermal stresses and deformations for some components. Additionally, too much heat was transferred through some of the walls. Moreover, a sealing of the hot box against the surrounding moderator and feedwater was missing. The design of the calandria tubes and the guide tubes for the control rods in the hot zone established a large heat transfer surface area, which was thought to influence the steam outlet temperature.

A third concept, introduced by Bittermann et al. [8], features three combined inlets and outlets using a coaxial pipe configuration. The inner pipe is connected with the hot box, which collects the supercritical steam and guides it outside the vessel. The outer pipe, which is the inlet nozzle in this case, is used to supply the core with feedwater and additionally works as a thermal sleeve to prevent contact between the outlet steam and the inner wall of the RPV. A spring at the end of the inner tube is proposed to compensate for thermal expansion of the hot box. Furthermore, piston ring seals are provided to prevent mixing of the inlet and outlet mass flow. In this conceptual design, the assembly of the internals resembles conventional PWR design, where the core rests on a support plate, which is connected to the core barrel. The barrel itself is suspended at the vessel flange. With this arrangement, the moderator tubes have to penetrate the steam plenum to reach the fuel assemblies below, creating several bypass-flows, which will reduce the outlet temperature of the core by mixing with the supercritical steam. Additonally, thermal stresses between the hot box, the outlet pipes and the RPV are not considered in this design. Hofmeister et al. [40] proposed a steam plenum which has been minimized in height to gain extra volume for an in-vessel coolant accumulator above the core. To avoid mixing of cold feed water and hot steam, C-ring seals are provided to prevent bypass flows between the head pieces of the fuel assembly clusters and the outside of the steam plenum.

The RPV design for the HPLWR has also been assessed for the application of important safety components in the case of transients. As recirculation pumps will not be required for the HPLWR concept, a postulated break of one of the inlet feedwater lines will cause an immediate flow reversal in the core and reduce the available water inventory in the vessel to cool the core. Investigations by Antoni and Dumaz [4] with the safety analysis code CATHARE showed that the cladding temperatures rise up to 1500 °C for the hottest assembly in the first 5 seconds of a loss of coolant accident in case of a feedwater line break. After that period, the automatic depressurization system (ADS) of the steam lines will cause subsequent flooding of the core and decrease the temperatures subsequently (Aksan et al. [1]). This temperature peak can be reduced significantly, if an additional safety component is installed in the feedwater lines to control and minimize this backflow, so that a reasonable amount of water is kept inside the vessel to cool the core. This component has to fulfill several requirements. It shall

- fit inside the inlet flange of the RPV to avoid damages,
- be a passive component without moving parts,
- have minimal pressure losses for regular operation condition,
- cause small flow rates i.e. high pressure losses in reverse direction,
- but react with a fast response in the case of transients.

The group of fluid diodes meets all those requirements and is chosen for a more detailed application analysis. Baker [5] executed a comparative study of three different types of diodes which work all on entirely different principles. These are the vortex diode, the scroll diode and the fluid flow rectifier. The performance has been evaluated using the ratio of the resistance coefficient in reverse flow direction to the resistance coefficient in regular operation condition. The results show, that the vortex diode gives the best performance with a ratio of 12.5:1 but can be as high as 50:1, followed by the fluid flow rectifier and the scroll diode. Furthermore, the vortex diode is less complex and has the smallest forward-flow loss coefficient. One drawback of the vortex diode is the overall size which is much larger for a given flow rate compared with the other two ones.

Several applications for the vortex diode as a safety component in nuclear reactors are known, e.g. in gas cooled British reactors (AGR) to control depressurization of the vessel in the event of a pipeline fracture (Syred and Roberts [85], King [48]). George et al. [34] developed a diode for the application in the primary circuit of such a reactor with high pressure differences of up to 4 MPa and temperatures around 330 °C. Owen and Motamed-Amini [67], [60] extended the investigation to diodes working with superheated steam. They found that the high resistance properties are significantly reduced for both directions in the transonic and supersonic regions. In this case, the critical mass flow for choking dominates the behavior of the diode.

Another application for the vortex diode is realized in the Japanese APWR to refill the reactor vessel after a LOCA and switch between a high flow rate in the beginning and a low flow rate for the core re-flooding (Ichimura et al. [43]). For the SWR 1000[7], passive outflow reducers are foreseen to significantly reduce the discharge mass flow from the RPV in the event of a postulated break in the emergency condenser return line (Brettschuh [10], Pasler [68]). A scaled outflow reducer has been investigated in experiments by Mueller [61] to achieve the characteristic of the component for different mass flows. The results indicated that the mass flow in the event of a pipe break is reduced to approximately 10 % of the value, which would result without the installed component in the RPV nozzles.

Review of the HPLWR Reactor Concept

The following conclusions can be drawn for the design of the pressure vessel and internals for the HPLWR using a single or a multi-pass core.

The coolant must be heated up in steps with intermediate mixing to keep the local hot channel temperature inside the material limit to achieve high core outlet temperatures around 500 °C. Two possible core concepts have been presented by Schulenberg et al. [76] for the HPLWR. The simpler approach is a two-step heat-up with intermediate mixing, while the three-step heat-up requires an even more complex core design. So far, only conceptual designs have been presented for this two pass and three pass core arrangement using the Hofmeister [40] fuel assembly cluster as a reference design. It has not been demonstrated yet that such a more sophisticated flow path is indeed mechanically feasible.

The higher temperature differences between the fuel assembly clusters imply thermal deformations, which have to be considered for the centering inside the core and the alignment of the single cluster with the control rods and the control rod guide assembly (CRGA). Buoyancy effects, which cause the cluster to swim up during operation have not yet been examined so far for the fuel assembly design by Hofmeister [40]. Due to the compact design of the cluster, the available space for a hold-down spring is very limited and conventional designs from PWR and BWR cannot be applied. To optimize the radial power profile and burn-up, it is desirable to use the same cluster for all different heat-up steps in the core, which has to be considered for the design of the head and foot piece of the cluster.

To reduce the non-uniformities of the coolant temperature between the heat-up steps, an upper and a lower mixing plenum have to be designed to provide intermediate mixing and to realize the sketched flow path of the two pass and three pass core. Both mixing chambers must be leak tight, to avoid addition of colder moderator water to the hot coolant. Moreover, the complete flow path of the coolant through the different fuel assembly clusters of the heat-up steps has to remain closed against leakage of colder moderator

---

[7]Projected advanced boiling water reactor by Areva NP.

water, even in case of larger thermal expansions of the components. To guarantee stable operation of the reactor, the different mass flows inside the core have to be adjustable, using orifices or openings, while having acceptable pressure losses along the flow path.

The design of the other internals and of the RPV can be deduced from state of the art technologies of pressurized water reactors. Components like the core barrel with its core support plate and alignments, the control rods and its guide tubes, the vessel closure head with its sealing, and the vessel bolt design have to be adapted to the higher pressure and design temperature, but feature a similar design. Other components have to be designed differently to account for the higher steam outlet temperature and design pressure. Thermal expansions between the internals and the RPV have to be controlled to minimize thermal stresses. Particularly the thermal displacements between the upper plenum, the core barrel and the RPV have to be minimized to allow a leak tight connection to the outlet flange. The application of a core with counter-current moderator and coolant mass flow also has an impact on the design of internals like core barrel and control rod guide assembly, which have to be modified to enable this complex flow path.

The operation conditions of 25 MPa and temperatures of up to 500 °C also influence the dimensioning of the RPV. In order to use conventional vessel materials, the vessel inner wall has to be kept in contact only with coolant at inlet temperature, requiring a thermal sleeve for the penetrating hot steam connection. As a consequence of the mentioned core design with several passes and the high design pressure, the diameter of the core and therefore the diameter of the RPV are quite large with wall thicknesses exceeding those of conventional PWR. The wall thickness of the vessel design has to be optimized to remain inside available forging limits.

In order to enhance the safety performance of the reactor in the case of a break of the feedwater line, the application of a vortex diode is required for the inlet of the RPV. Those passive safety devices have already be considered for advanced boiling water reactors but only for much smaller flow rates and pressure differences. Starting with a simple vortex design, the component has to be optimized to have a low flow resistance for the normal operation condition. Additionally, the limited space inside the inlet flange of the RPV has to be considered for the design. To evaluate the performance of the optimized design, the flow characteristic has to be predicted for both directions. The resulting behavior is needed as an input for safety system analyses.

1.2 Aim and Outline of the Study

Besides the higher pressure and higher steam temperature, the design concept of a SCWR differs significantly from a conventional LWR by a different core concept. Therefore, the aim of this thesis is the design of the RPV and its internals for a supercritical water-cooled reactor for different core arrangements. Based on a first design of a fuel assembly cluster for a HPLWR with a single pass core, the surrounding internals and the RPV are dimensioned for the first time, following the safety standards of the nuclear safety standards commission in Germany. Furthermore, this design is extended to the incorporation of core arrangements with two and three passes. For those concepts, the fuel assembly cluster and the internals are redesigned to facilitate the complex flow path for the multi-pass concepts. The design of the internals and of the RPV are verified using mechanical or, in the case of large thermal deformations, combined mechanical and thermal stress analyses.

Additionally, a passive safety component for the feedwater inlet of the RPV of the HPLWR is designed. Its purpose is the reduction of the mass flow rate in case of a LOCA for a feedwater line break. Starting with a simple vortex diode, several steps are executed to enhance the performance of the diode and adapt it to this application. Then, this first design is further optimized using combined 1D and 3D flow analyses. Parametric studies determine the performance and characteristic for changing mass flow rates for this backflow limiter. To meet these objectives, the following design steps and analyses are performed.

The detailed design of the different assembly clusters with the key components head and foot piece is described in chapter 7. Based on the design of the fuel assembly cluster for the one pass core, design modifications are introduced to obtain the designated flow path through the assemblies for the two and three pass core arrangement. An optimized design for the carrier structure of the fuel rods is presented to reduce the overall coolant pressure drop for the several passes through the core. Additionally, the sealing concept with C-ring seals and piston rings is specified, which ensures leak tightness for the whole coolant flow path.

The internals are dealt with in the following chapter 8, where the design of the core barrel with core support plate and lower plenum, the steel reflector, the upper plenum (steam plenum) and the control rod guide assembly are explained in detail. Starting with the incorporation of the core into the core barrel, the positioning and alignment of all internals into the RPV is described. Special attention is payed to the fixation of the steam plenum, which rests on support brackets that are attached to the circumference of the vessel. The rather unique design decouples thermal displacements between the internals and the RPV, which minimizes occurring thermal stresses. It also allows for a leak tight connection between the hot tube and the steam plenum. Furthermore, design details of the lower and upper plenum to allow a leak tight coolant flow path between the different heat-up steps are discussed.

The design of the RPV and its closure head with the connecting nuts and bolts is referred to in chapter 9. The three pass core vessel is explained in detail since it experiences the highest loads of all three investigated designs. Accordingly, the function and design of the thermal sleeve for the hot pipe is described together with its implementation into the outlet flange of the RPV. The reactor design with RPV and internals for all three presented core concepts is summarized in chapter 6. Here, the dimensions of the RPV for all three different core concepts are given together with the specific core arrangement.

The safety standards of the nuclear safety standards commission (KTA) in Germany, which are discussed in chapter 2, are used for the dimensioning of the components. A summary of the considered load cases and the performed mechanical analyses for the RPV and the internals is given in chapter 10. Additionally, the chapter includes a tentative material selection for the components to obtain material strength characteristics for the dimensioning.

For the optimized design of the RPV and its outlets and for the steam plenum, a coupled thermo-mechanical finite element analysis is performed. A brief overview for the finite element method and its application is described in chapter 4. The thermal loads are pre-calculated for both analyses using heat transfer correlations as described in chapter 2. The evaluation of the resulting deformations and stresses is discussed in chapter 11.

Chapter 12 describes the fluidic optimization of the backflow limiter. In a first op-

1.2 Aim and Outline of the Study

timization step, the general configuration for the application in the inlet flange of the RPV is determined using 1D flow analysis, as described in chapter 3. This first design is further optimized using combined 1D and 3D CFD flow analyses in a second optimization step. The applied computational fluid dynamics method is described in chapter 5, together with a verification and validation of the applied turbulence model and grid. Parametric studies are executed and described at the end of chapter 12 to determine the characteristic of the backflow limiter.

A summary of the presented design for the RPV and the internals for all three core concepts is given in chapter 13. Finally, conclusions in relation to the HPLWR project are drawn with an outline of further work.

Part I
Fundamental Design Methods

2 Mechanical Analysis of the Vessel and its Internals

The pressure vessel as well as its internals including the core barrel, steam plenum and control rod guide tubes, are subject to a number of thermal, mechanical and irradiation constraints. Therefore, mechanical analyses need to be performed for critical cross sections of each component under mechanical and thermal loadings, corrosion, erosion, and irradiation. Mechanical and thermal boundary conditions directly influence the components life time. They may also have indirect effect; for example different coolant temperatures might cause temperature gradients inside the component structure and lead to different thermal expansion of the components. Corrosion and erosion may lead to wall-thinning and in connection with stresses to cracking. Irradiation causes embrittlement of the vessel material in the core area and causes additional heat sources. However, since there are no final material investigation results for corrosion, erosion, and irradiation effects for the chosen materials in relation to supercritical water, only mechanical and thermal loadings are considered for the analyses here. These loadings include forces and moments, imposed deformations and temperature gradients caused by the fluid, the component itself, by adjacent components, and by transferred loadings.

2.1 Applied Safety Standards

Like with ordinary pressure vessels, certain safety criteria have to be considered for the construction and operation of nuclear reactors. According to the Atomic Energy Act these requirements shall be met in accordance with the state of science and technology. In Germany, the safety standards of the Nuclear Safety Standards Commission (KTA) have the task of specifying those safety related requirements. Since the focus of the study lies on the design and analysis, two safety standards are of special interest in this case. The KTA standard 3201.2 [52] deals with the components of the reactor coolant pressure boundary of light water reactors, which includes the design and analysis of the pressure vessel, the feedwater inlets, the main steam outlet nozzles and the closure head. The reactor pressure vessel internals such as the core barrel or the control rod guide tubes are dealt with in KTA 3204 [50]. Each safety standard is specified using a classification code which includes load case classes, design and analysis, calculation procedures and design principles to satisfy the required quality.

2.2 Load Case Classes

The load case classes include all possible conditions and changes to the system in connection with the specified loading level acting on the components. These loading levels refer to allowable loadings. The load case classes can be split up into six different categories:

- design load cases (AF), which cover normal operational load cases causing maximum primary stresses in the component,

- normal operational load cases (NB), which include operating conditions and changes to the system for functionally fit conditions as start-up, full-load operation and shutdown,

- anomalous operational load cases (AB), which refer to deviations from the normal operating load cases such as functional disturbance or control errors,

- test load cases (PF), which include the first pressure test as well as periodic tests,

- incidents like emergencies (NF) and accidents (SF).

For the present analysis only the following loading levels are considered. The various loading levels are specific for each component and the loading limits are determined in a way, that the integrity of the component is ensured at any loading level.

- Design loading (level 0) includes the maximum design pressure, design temperature and additional design loads which occur for full-load operation for each component,

- level A service limits refers to loadings from normal operational load cases (NB) where primary stresses and the superposition of different stress categories (e.g. primary and secondary stresses) should not exceed the permitted values.

Table 2.1 gives a summary of component loadings and their classification into loading levels for the pressure vessel, feedwater inlet, main steam outlet nozzles, and the closure head. The steam plenum, which is dealt with in KTA 3204 follows the same outline. All other components are evaluated using the maximum primary stresses only.

Service loading level	Static loadings							Transient loadings
	Design pressure	Design temperature	Pressure	Temperature	Deadweight and other loads	Mechanical loads, reaction forces	Restraint to thermal expansion	Transient loads (pressure, temperature mechanical loads), dynamic loading
Level 0	×	×			×			
Level A			×	×	×	×	×	×

Table 2.1: Classification of the different loadings and their superposition into service level 0 and A.

2.3 Design Requirements

The design of all components has to meet specific functional requirements and shall not lead to an increase of loadings and stresses. Especially in areas of structural discontinuity, for example in the vicinity of the main steam outlet nozzles, it is desired to have a favorable stress distribution. Another example would be the wall thickness transition from the lower spherical vessel bottom to the thicker cylindrical part, where a favorable distribution of stresses is desired. This is also important for transient temperature loadings where abrupt changes in wall thickness will cause higher local stresses. Furthermore, welds have to be avoided in areas of high local stresses and irradiation. This is very important at the circumferential shell of the vessel surrounding the core, where the γ-irradiation can damage those welds. Additionally, all selected materials for the components have to meet specific requirements, which include the strength, ductility, physical properties like the coefficient for thermal expansion, and corrosion resistance. Those requirements are specified in KTA 3201.1 [51]. The design has to be maintenance friendly, meaning that for example adequate accessibility for a fast loading and unloading procedure of spent fuel assembly clusters is possible. The mentioned measures are valid for each component and have to be applied using the corresponding safety standard.

2.4 Dimensioning

Dimensioning shall be based on the basis of the design loading level (level 0) and take into account the loadings and service limits of level A, as far as these concern dimensioning. Dimensioning rules are applied for the reactor pressure vessel, the main steam outlet nozzles, the feedwater inlet, the closure head and the steam plenum. State-of-the-art techniques which are used to verify the dimensioning are finite elements analyses with the software ANSYS as described in chapter 4.

2.5 Mechanical Behavior

The analysis of the mechanical behavior has to verify that the components are capable of withstanding all loadings according to the loading levels.

2.5.1 Stress Analysis

By means of a stress analysis along with a classification of stresses and limitation of stress intensities it shall be proved, that no inadmissible distortions and only limited plastic deformations occur. The classification of stresses depends on the cause of stress and its effect on the mechanical behavior of the structure. Three different classes can be defined:

- Primary stresses are stresses which satisfy the laws of equilibrium of external forces and momentum, and which are not self-limiting concerning the mechanical behavior. Two different categories of primary stresses exist, the membrane stresses (P_m) and bending stresses (P_b). Membrane stresses are defined as the average value of the stress component distributed across the thickness. Bending stresses are stresses that can be altered linearly across the considered section or thickness and proportionally to the distance from the neutral axis.

- Secondary stresses (Q) are stresses developed by constraints due to geometric discontinuities and by the use of materials of different elastic moduli under external loads, and by constraints due to differential thermal expansions. Secondary stresses are self-limiting as they lead to plastic deformation when equalizing different local distortions. Only stresses that are distributed linearly across the cross section are considered to be secondary stresses.
- Peak stresses (F) are the increment of stress which are additive to the respective primary and secondary stresses. They do not cause any noticeable distortion and are only important to fatigue fracture in conjunction with primary and secondary stresses. Peak stresses comprise deviations from nominal stresses at notches due to pressure and temperature.

The evaluation and superposition of stresses are carried out for each load case where the stresses acting in the same direction are added separately or for different stress categories jointly. The allowable values for stress intensities and equivalent stress ranges for the linear-elastic analysis of the mechanical behavior for the analyzed loading levels 0 and A can be seen in Table 2.2.

	Stress category	Level 0	Level A
Primary stresses	P_m (I)	S_m	-
	$P_m + P_b$ (II)	$1.5 \cdot S_m$	-
Primary plus secondary stresses	$P_m + P_b + Q$ (III)	-	$3 \cdot S_m$
Primary plus secondary stresses plus peak stresses	$P_m + P_b + Q + F$ (IV)	-	$D \leq 1.0$; $2 \cdot S_a$

Table 2.2: Allowable values for stress intensities and equivalent stress ranges for a linear-elastic analysis of the mechanical behavior for service level 0 and A for the components pressure vessel, feedwater inlet, main steam outlet nozzles, closure head, and the steam plenum.

The determination of stress intensities and equivalent stress ranges is based on the stress theory of von Mises which takes, for a three-dimensional set of coordinates x, y, and z, the following form:

$$\sigma_{V,v.\,Mises} = \sqrt{\sigma_x^2 + \sigma_y^2 + \sigma_z^2 - (\sigma_x \cdot \sigma_y + \sigma_x \cdot \sigma_z + \sigma_y \cdot \sigma_z) + 3 \cdot \left(\tau_{xy}^2 + \tau_{xz}^2 + \tau_{yz}^2\right)} \quad (2.1)$$

where the algebraic sums of all normal stresses σ [Nmm^{-2}] and shear stresses τ [Nmm^{-2}] acting simultaneously are calculated for the general primary membrane stresses or the sum of primary bending stresses and the general primary membrane stresses. The stress intensities and equivalent stress ranges for each service loading level are limited in dependency of the mechanical behavior of the material. The limitation is based on the stress intensity factor S_m for primary and secondary stresses, and on the cumulative damage factor D for the peak stresses. The S_m value is obtained on the basis of the temperature T of the respective component. For the design level 0, however, the design temperature shall be used. For ferritic and austenitic materials the S_m value is derived as follows:

$$S_m = \min \cdot \left\{ \frac{R_{p0.2T}}{1.5}; \frac{R_{mT}}{2.7}; \frac{R_{mRT}}{3} \right\} \quad (2.2)$$

2.6 Thermal Loads

with the 0.2 % elevated temperature proof stress $R_{p0.2T}$ in Nmm^{-2}, the minimum tensile stress at elevated temperature R_{mT} in Nmm^{-2}, and the minimum tensile stress at room temperature R_{mRT} in Nmm^{-2}. The minimum values are taken from KTA 3201.1 [51] for the respective materials of the components.

2.5.2 Fatigue Analysis

For cyclic loading of a component, an elastic fatigue analysis is performed to avoid fatigue failure. This evaluation method is based on a linear-elastic stress strain relationship, such that the equivalent stress range resulting from primary and secondary stresses does not exceed a value of $3 \cdot S_m$. Peak stresses in level A are evaluated the following way. For each cycle, the number of cycles to failure, n, at a given stress amplitude $S_a = \frac{1}{2} \cdot (P_m + P_b + Q + F)$ is taken from the fatigue curves in Appendix A.1.1 or Appendix A.1.2, respectively, and compared with the allowable number of cycles $\hat{n}$. The total cumulative damage factor

$$D = \sum \frac{n}{\hat{n}} \qquad (2.3)$$

shall not exceed D=1. A safety margin shall be included in $\hat{n}$, accounting also for embrittlement of the material.

2.5.3 Thermal Strain Analysis

Due to the different temperatures of the fluid inside the reactor, the components have different thermal expansions. To determine the linear thermal expansion of each component, the following equation according to Dubbel [21] is used:

$$l = l_0 \left(1 + \alpha_T (t - t_0)\right) \qquad (2.4)$$

where α_T is defined as:

$$\alpha_T = \frac{l - l_0}{l_0 \cdot (t - 0)^\circ C} \qquad (2.5)$$

Here, the linear thermal expansion for solid materials for a given temperature range can be deduced from [21].

2.6 Thermal Loads

Thermal loads occur due to a heat flux caused by the fluid which causes temperature differences in the component. For the calculations, heat transfer coefficients between the fluid and the surface of the component as well as fluid temperatures are needed to define these thermal loads.

2.6.1 Dimensionless Numbers

Heat transfer is described using the following dimensionless numbers:

The Prandtl number is approximating the ratio of momentum diffusivity (viscosity) and thermal diffusivity using the following equation:

$$\Pr = \frac{\nu}{a} \qquad (2.6)$$

where ν is the kinematic viscosity in m²s⁻¹, and a is the temperature diffusivity of the fluid in m²s⁻¹.

The Reynolds number Re is used to identify the flow regime of the different flows in the reactor. It is defined as the ratio of inertial forces to viscous forces:

$$\mathrm{Re} = \frac{v_s \cdot L}{\nu} \qquad (2.7)$$

where v_s is the mean fluid velocity in ms⁻¹, L is the characteristic length in m (usually the hydraulic diameter), and ν is the kinematic viscosity of the fluid in m²s⁻¹.

The Nusselt number is used to measure the enhancement of heat transfer due to convection. It relates the convective heat transfer to the conductive heat transfer in perpendicular to the flow direction:

$$Nu = \frac{\alpha \cdot L}{\lambda} \qquad (2.8)$$

where λ is the thermal conductivity of the fluid in Wm⁻¹K⁻¹, and α is the heat transfer coefficient in Wm⁻²K⁻¹.

The Grashof number describes the ratio of the buoyancy to the viscous forces acting on a fluid, which in the case of natural convective mass transfer, is defined as:

$$Gr = \frac{g \cdot l^3}{\nu^2} \cdot \beta \cdot \frac{q}{\alpha} \qquad (2.9)$$

where g is the acceleration due to Earth's gravity with 9.81 ms⁻², l is the characteristic length responsible for buoyancy effects in m (usually the volume height), ν is the kinematic viscosity of the fluid in m²s⁻¹, β is the volumetric thermal expansion coefficient in K⁻¹, q is the heat flux between the fluid and the surface, and α is the heat transfer coefficient of the fluid in Wm⁻²K⁻¹.

Some correlations use the Rayleigh number instead. It can be expressed as:

$$Ra = Gr_c \cdot \Pr \qquad (2.10)$$

2.6.2 Overall Heat Transfer

Heat transition is defined as the stationary transport of heat through a one or multi-layer wall with heat transfer at both surfaces. The local heat transfer coefficient α of forced convection is depending on the flow length, the temperature of the fluid, and the temperature of the wall. The heat flux q between the fluid and the wall is defined as:

$$q = \alpha \cdot (T_{fluid} - T_{wall}) \qquad (2.11)$$

2.6 Thermal Loads

where q is given in Wm^{-2}. To determine the heat transition coefficient k for the heat transfer through a cylindrical circular tube with the wall thickness δ_{wall}, the following equation can be used [91]:

$$\frac{1}{k \cdot A} = \frac{1}{\alpha_1 \cdot A_1} + \frac{\delta_{wall}}{\lambda_{wall} \cdot A_{wall}} + \frac{1}{\alpha_2 \cdot A_2} \quad (2.12)$$

where A is an arbitrary chosen transfer area in m^2, α_1 and α_2 are the heat transfer coefficients from fluid one and fluid two to the corresponding wall in Wm^{-2}K^{-1}, A_1 and A_2 are the transfer areas of the corresponding fluid in m^2. The mean transfer area for the tube wall, A_{wall} is defined as:

$$A_{wall} = \frac{A_1 - A_2}{\ln \frac{A_1}{A_2}} \quad (2.13)$$

2.6.3 Heat Transfer for Concentric Annular Gaps

For a fully developed turbulent flow through a pipe of length l_{tube} (Re $\geq 10^4$), the following equation by Gnielinski [36] can be applied to determine the Nusselt number:

$$Nu_{pipe} = \frac{\left(\frac{\xi}{8}\right) \cdot Re \cdot Pr}{1 + 12.7 \cdot \sqrt{\frac{\xi}{8}} \cdot \left(Pr^{2/3} - 1\right)} \left[1 + \left(\frac{d_i}{l_{tube}}\right)^{2/3}\right] \quad (2.14)$$

with the Konakov relation for the pressure loss coefficient [49]:

$$\xi = (1.8 \cdot \log_{10} Re - 1.5)^{-2} \quad (2.15)$$

for non-circular pipes the inner diameter d_i in equation 2.14 is substituted by the hydraulic diameter d_h:

$$d_h = \frac{4 \cdot F}{U} \quad (2.16)$$

where F is the flow cross section area and U the inner perimeter of the non-circular pipe. The heat transfer of fluids with different temperatures is influenced by the direction of the heat flux (cooling or heating of the wall). For turbulent flow, Hufschmidt and Brueck [42] introduce the following factor to take the direction of the heat flux into account, which is based on the flow viscosity relationship from Sieder and Tate [80]:

$$\left(\frac{Pr}{Pr_w}\right)^{0.11} \quad (2.17)$$

where Pr is the Prandtl number with the corresponding properties for the average temperature of the fluid, and Pr_w is the Prandtl number for the properties of the fluid at wall temperature. Therefore, the iteratively determined Nusselt number Nu is defined as:

$$Nu = Nu_{m,T} \cdot \left(\frac{Pr}{Pr_w}\right)^{0.11} \quad (2.18)$$

where $Nu_{m,T}$ is the Nusselt number with the corresponding properties for the average temperature of the fluid. The heat transfer coefficient α is then determined using the following equation:

$$\alpha = \frac{Nu \cdot \lambda}{d_h} \qquad (2.19)$$

where λ is the thermal conductivity of the fluid in $Wm^{-1}K^{-1}$.

Two different boundary conditions can be distinguished for the application in the reactor. The first case (A) considers heat transfer from the inner pipe to the fluid, where the outer pipe is isolated. Petukhov and Roizen [70] found the following relation for turbulent flow in a concentric annular gap:

$$\frac{Nu_{m,T,A}}{Nu_{pipe}} = 0.86 \cdot \left(\frac{d_i}{d_o}\right)^{-0.16} \qquad (2.20)$$

where the hydraulic diameter for concentric pipe flow d_h is written as $d_o - d_i$, $Nu_{m,T}$ is the mean Nusselt number for the average temperature of the fluid, and Nu_{pipe} is the general Nusselt number for turbulent pipe flow according to Equation 2.6.3. For the second case (B) the heat transfer of the outer pipe to the fluid is considered, with an isolated inner pipe. According to [70] the following correlation can be used:

$$\frac{Nu_{m,T,B}}{Nu_{pipe}} = 1 - 0.14 \cdot \left(\frac{d_i}{d_o}\right)^{0.6} \qquad (2.21)$$

Both equations are valid for $0 \leq \frac{d_i}{d_o} \leq 1$.

2.6.4 Heat Transfer for Free Convection on Vertical Surfaces

The mean heat transfer coefficient for the laminar and turbulent flow condition for $10^{-1} \leq Ra \leq 10^{12}$ is defined with the dimensionless Nusselt number according to Churchill and Chu [18]:

$$Nu = \left(0.825 + 0.387 \left(Ra \cdot f_1\left(Pr\right)\right)^{\frac{1}{6}}\right)^2 \qquad (2.22)$$

In this case, the characteristic length l (according to Equation 2.9) is the height of the vertical wall, where the function $f_1(Pr)$ includes the influence of the Prandtl number in the range of $0.001 < Pr < \infty$:

$$f_1\left(Pr\right) = \left(1 + \left(\frac{0.492}{Pr}\right)^{\frac{9}{16}}\right)^{-\frac{16}{9}} \qquad (2.23)$$

3 Flow Analysis of a Backflow Limiter

For pre-dimensioning of the backflow limiter, a 1D flow analysis is performed. Applying the steady state 1D momentum equation for incompressible fluids in accordance with the conservation of angular momentum the velocities and pressures for certain positions along the flow path inside the backflow limiter are determined (see Figure 3.1). The Bernoulli-equation according to [62] is used to calculate the required parameters.

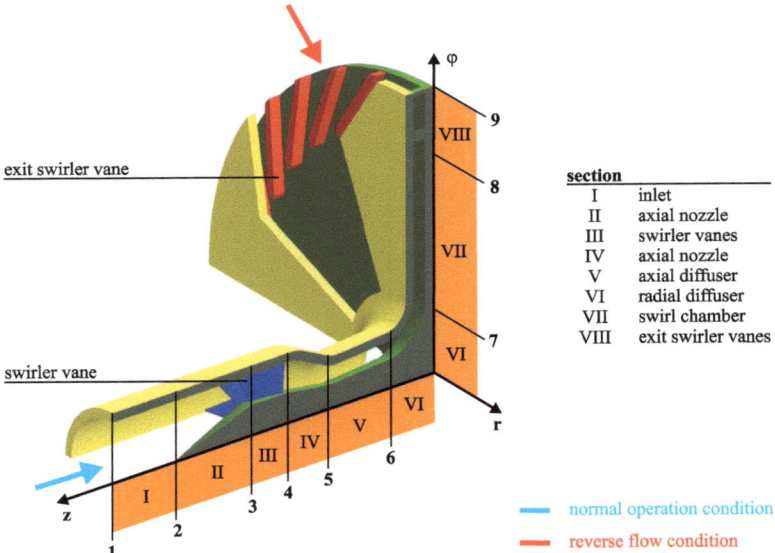

Figure 3.1: Quarter section of the backflow limiter with the several sections and positions for the 1D analysis.

The backflow limiter is divided into the following flow sections: the inlet section (I), inlet nozzle (II), inlet swirler vanes (III), second axial nozzle (IV), axial diffuser (V), radial diffuser (VI), swirl chamber (VII) and exit swirler vanes (VIII). To determine the pressure loss for each section, the static pressures and the dynamic pressures are calculated for each position between the different sectors marked in Figure 3.1 with position numbers 1 to 9. The flow for the normal operation condition is marked with a blue arrow, the reverse flow direction is indicated in red.

3.1 Performance Factor

To evaluate the performance of a backflow limiter for different operational conditions the definition by Baker [5] is applied. Here the performance Σ is defined as the ratio of the resistance coefficient K_A in reverse flow direction to the resistance coefficient K_B in the regular flow condition under steady-state conditions at the same Reynolds number:

$$\Sigma = \frac{K_A}{K_B} \quad (3.1)$$

where the resistance coefficient K is defined as the ratio of the pressure loss measured across the backflow limiter to the mean velocity in the pipe at the entry of the backflow limiter:

$$K = \frac{2 \cdot \Delta p}{\rho \cdot \bar{v}^2} \quad (3.2)$$

The mean fluid velocity $\bar{v}$ is defined as the average axial velocity over the flow cross section area of the pipe inlet. The Reynolds number as defined in Equation 2.7 is used for the dimensionless comparison of the regular flow condition and the reverse flow condition.

3.2 Pressure Loss Coefficients

Due to the varying flow cross sections within the backflow limiter, it is divided into n sections for which the local pressure loss is determined and added up to the overall pressure loss Δp of the component.

$$\Delta p = \frac{\rho}{2} \cdot \sum_{I}^{n} \zeta_i \cdot \bar{v}_i^2 \quad (3.3)$$

where ζ_i is the individual pressure loss coefficient for each section i, ρ is the density of the coolant, $\bar{v}_i$ is the fluid velocity at the inlet of section i. For a first calculation, pressure loss coefficients are estimated using general friction factors used for turbo engine design (Sigloch [81]) for each flow cross section.

3.3 Swirl

To evaluate the swirl distribution inside the backflow limiter along the flow path, the following general approach is used to calculate the swirl S for a cylindrical coordinate system.

$$S = \rho \cdot r \times v_{peripheral} \cdot dV \quad (3.4)$$

where ρ is the density, r is the radius in ϕ-direction, and $v_{peripheral}$ is the peripheral velocity of the fluid element which has the volume dV.

Part II
Numerical Methods

4 Finite Element Simulation

The finite elements method implemented in ANSYS WORKBENCH/CLASSIC is used to verify the structural integrity of critical components as defined by the mechanical analysis described in chapter 2. Two components with very complex geometries and high mechanical and thermal loads are determined for the verification. This is, on one hand, the reactor pressure vessel, which experiences a high pressure difference from the inside to the outside of more than 25 MPa, and on the other hand, the steam plenum with temperature differences throughout the material as high as 200 °C.

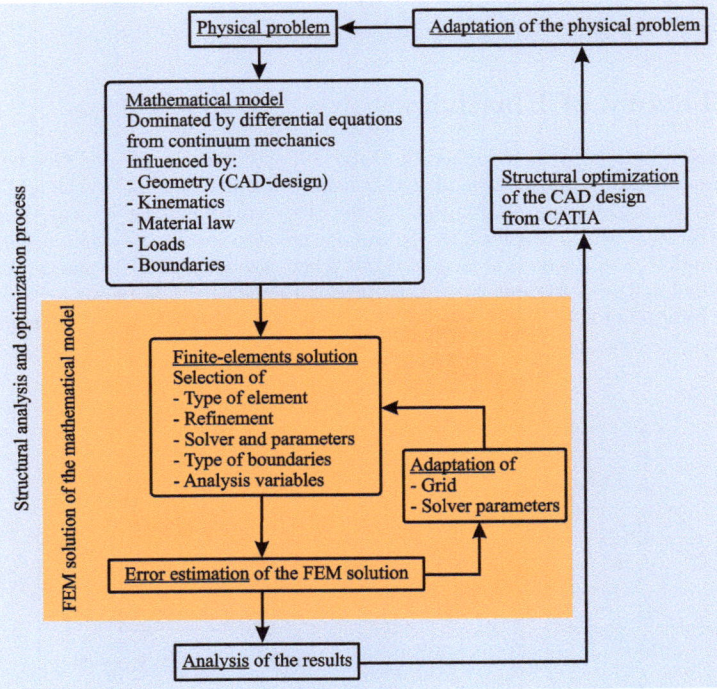

Figure 4.1: Calculation and optimization scheme for the finite elements analysis of investigated components according to Müller and Groth [58].

The method allows to resolve, even under highly complex loads, the local stresses and deformations to perform a structural optimization of the component. Additionally, mechanical and thermal loads can be applied simultaneously.

Figure 4.1 shows the general analysis process (Müller and Groth [58]) for the reactor pressure vessel and the steam plenum using the finite elements method that is implemented in the code ANSYS CLASSIC/WORKBENCH. The idealized physical problem is expressed through a set of differential equations representing the mathematical model. Assumptions about the geometry, kinematic relationships, the material law and the applied loads and boundaries influence this model and the results. The obtained numerical solution is also dependent on the type of elements that are applied, the refinement of the mesh and the employed solver parameters. Two different steps can be observed in the chart. The first step, marked with an orange background in Figure 4.1 involves an iterative process to control the accuracy of the finite elements solution for the mathematical model. This is done by adapting and increasing the local mesh refinement in regions where high gradients occur until the difference between two successive solutions satisfies the defined deviation range. The second step is marked in a blue background in Figure 4.1 and evaluates the results from the analysis in order to perform a structural optimization of the design, which in turn has an impact on the physical problem and the mathematical model.

4.1 Theory of Elasticity

The theory of elasticity as described in Szabó [86] and Timoshenko and Goodier [87] is used to mechanically analyze structures that behave in a linearly elastic fashion. The state of stress and deformation distribution for the observed component is depending on an equation system, which is defined by the equilibrium stress state and strain-displacement for the volume, and the general form of Hooke's law, also referred to as constitutive form. The normal and shear components of the three dimensional stress tensor for an element volume according to Schnell et al. [73] are shown in Figure 4.2, to the left. In equilibrium $\tau_{yx} = \tau_{xy}$, $\tau_{zx} = \tau_{xz}$, $\tau_{zy} = \tau_{yz}$, and the stress tensor matrix is symmetric.

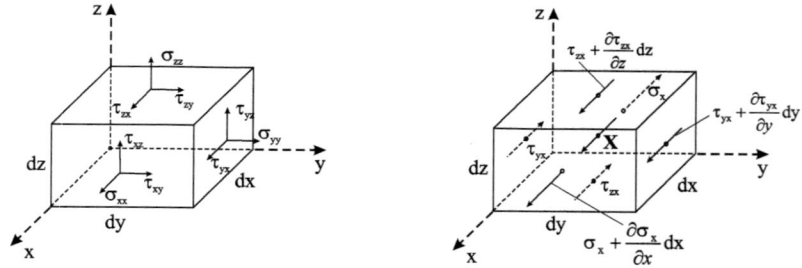

Figure 4.2: Normal and shear components of the three dimensional stress tensor for an element volume according to [73] (left side), occurring forces **X** and stresses to satisfy equilibrium condition for the zy-plane of the element volume (right side).

The three dimensional strain-displacement equations can be considered as a superposition of two effects: stretching in direction of the load and shrinking caused by the load in

4.1 Theory of Elasticity

perpendicular direction. For small deformations and therefore respective displacements d_i in the x, y, and z directions, the strain ε is defined as:

$$\varepsilon = \varepsilon_x + \varepsilon_y + \varepsilon_z = \frac{\partial d_x}{\partial x} + \frac{\partial d_y}{\partial y} + \frac{\partial d_z}{\partial z} \tag{4.1}$$

and the shear rate γ is defined as:

$$\gamma_{xy} = \frac{\partial d_y}{\partial x} + \frac{\partial d_x}{\partial y}, \quad \gamma_{xz} = \frac{\partial d_z}{\partial x} + \frac{\partial d_x}{\partial z}, \quad \gamma_{yz} = \frac{\partial d_z}{\partial y} + \frac{\partial d_y}{\partial z} \tag{4.2}$$

The state of stress as defined by the stress tensor is at equilibrium state if the following conditions are satisfied:

$$\frac{\partial \sigma_x}{\partial x} + \frac{\partial \tau_{yx}}{\partial y} + \frac{\partial \tau_{zx}}{\partial z} + X = 0,$$

$$\frac{\partial \tau_{xy}}{\partial x} + \frac{\partial \sigma_y}{\partial y} + \frac{\partial \tau_{zy}}{\partial z} + Y = 0, \tag{4.3}$$

$$\frac{\partial \tau_{xz}}{\partial x} + \frac{\partial \tau_{yz}}{\partial y} + \frac{\partial \sigma_z}{\partial z} + Z = 0,$$

where the stress forces σ and τ must be balanced by the body forces X, Y, and Z (gravitation, electromagnetic forces, etc.) on the cube, yielding the equilibrium conditions. Figure 4.2 gives an example for the balancing of the body force X in x-direction and the corresponding stress forces.
For the closure of the equation system the general form of Hooke's law [86], [87] for an isotropic and homogeneous material with temperature gradients is needed:

$$\varepsilon_x = \frac{1}{E}\left(\sigma_x - \mu\left(\sigma_y + \sigma_z\right)\right) + \alpha_T \Delta T,$$

$$\varepsilon_y = \frac{1}{E}\left(\sigma_y - \mu\left(\sigma_z + \sigma_x\right)\right) + \alpha_T \Delta T, \tag{4.4}$$

$$\varepsilon_z = \frac{1}{E}\left(\sigma_z - \mu\left(\sigma_x + \sigma_y\right)\right) + \alpha_T \Delta T,$$

$$\gamma_{xy} = \frac{\tau_{xy}}{G}, \quad \gamma_{xz} = \frac{\tau_{xz}}{G}, \quad \gamma_{yz} = \frac{\tau_{yz}}{G}. \tag{4.5}$$

with α_T being the thermal expansion coefficient, ΔT the occurring temperature difference, and G being the shear modulus which can be expressed in terms of the Young's modulus E as $G = \frac{E}{2(1+\mu)}$. For those 15 equations with 15 variables the differential equations can be expressed in terms of the unknown displacements Δd_i (i=x,y,z) as:

$$G\left(\Delta d_x + \frac{1}{1-2\mu}\frac{\partial \varepsilon}{\partial x}\right) + X = 0,$$

$$G\left(\Delta d_y + \frac{1}{1-2\mu}\frac{\partial \varepsilon}{\partial y}\right) + Y = 0, \quad (4.6)$$

$$G\left(\Delta d_z + \frac{1}{1-2\mu}\frac{\partial \varepsilon}{\partial z}\right) + Z = 0,$$

with $\Delta d_i = \frac{\partial^2 d_i}{\partial x^2} + \frac{\partial^2 d_i}{\partial y^2} + \frac{\partial^2 d_i}{\partial z^2}$. Those differential equations can be applied for problems, where the boundary conditions include prescribed displacements. They can be transformed into equations for the unknown stresses $\sigma_{x,y,z}$ by eliminating the displacements and their deviations:

$$\Delta\sigma_x + \frac{1}{1+\mu}\frac{\partial^2 \sigma}{\partial x^2} + 2\frac{\partial X}{\partial x} + \frac{\mu}{1-\mu}\left(\frac{\partial X}{\partial x} + \frac{\partial Y}{\partial y} + \frac{\partial Z}{\partial z}\right) = 0,$$

$$\Delta\sigma_y + \frac{1}{1+\mu}\frac{\partial^2 \sigma}{\partial y^2} + 2\frac{\partial Y}{\partial y} + \frac{\mu}{1-\mu}\left(\frac{\partial X}{\partial x} + \frac{\partial Y}{\partial y} + \frac{\partial Z}{\partial z}\right) = 0, \quad (4.7)$$

$$\Delta\sigma_z + \frac{1}{1+\mu}\frac{\partial^2 \sigma}{\partial z^2} + 2\frac{\partial Z}{\partial z} + \frac{\mu}{1-\mu}\left(\frac{\partial X}{\partial x} + \frac{\partial Y}{\partial y} + \frac{\partial Z}{\partial z}\right) = 0.$$

where $\sigma = \sigma_x + \sigma_y + \sigma_z$. The shear stresses $\tau_{xy,xz,yz}$ can be expressed by:

$$\Delta\tau_{xy} + \frac{1}{1+\nu}\frac{\partial^2 \sigma}{\partial x \partial y} + \frac{\partial X}{\partial y} + \frac{\partial Y}{\partial x} = 0,$$

$$\Delta\tau_{xz} + \frac{1}{1+\nu}\frac{\partial^2 \sigma}{\partial x \partial z} + \frac{\partial X}{\partial z} + \frac{\partial Z}{\partial x} = 0, \quad (4.8)$$

$$\Delta\tau_{yz} + \frac{1}{1+\nu}\frac{\partial^2 \sigma}{\partial y \partial z} + \frac{\partial Y}{\partial z} + \frac{\partial Z}{\partial y} = 0.$$

This form of the differential equations needs prescribed stresses as boundary condition in order to solve them. If a mixed form of both boundary conditions (displacements and stresses) is used, both equation systems have to be applied.

4.2 Discretization Method

The finite element method allows the calculation of coefficients influencing the mechanical stress and deformation behavior of the analyzed component. According to the finite element method, the geometry under investigation is divided into a set of discrete volumes or finite elements interconnected by points called nodes. The elements are generally unstructured (Bathe [6]), giving them the ability to deal with arbitrary geometries and allowing easy refinement. In the case of three dimensional structures, tetrahedral and hexahedral elements are used to model the geometry. The continuum formulation for each element is coupled to the attached elements via the numerical nodes at the element

4.2 Discretization Method

corners. The equation system is influenced by the imposed mechanical and thermal constraints and initial conditions. The so-called principle of virtual displacements uses the local node displacements as unknown parameters for the analysis, in the case of linear material behavior, the virtual work for each element is determined to solve the problem (Gallagher [32]). This is also the suggested method by the KTA ([52], [50] clause B.3). To receive the individual displacements, the system stiffness matrix is set up by using the individual displacement of each node together with the general form of Hooke's law (see Equations 4.4, 4.5). Using this matrix and the equilibrium conditions for each node results in an equation system for the unknown displacements, which is solved in several iteration steps using the appropriate boundary conditions.

For the simplest elements, displacements are linearly approximated for the element boundary and the inside of the element. For a planar element, the occurring stress and shear tensors can be seen in Figure 4.3 on the left side, while the displacement of one corner node of a planar triangle element is shown on the right side.

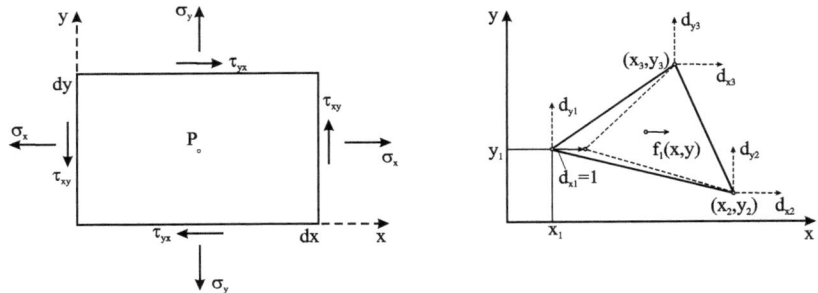

Figure 4.3: Planar stress state with displayed stress and shear tensors (left side) and planar triangle element with a unit displacement state $d_{x1} = 1$ (right side).

For the unit displacement state, $d_{x1} = 1$ and $d_{y1} = 1$, the displacement function $f_1(x,y)$ can be written as:

$$f_1(x,y) = \frac{1}{2 \cdot A_{element}} \left(x(y_3 - y_2) + y(x_2 - x_3) + x_3 y_2 - x_2 y_3 \right) \quad (4.9)$$

where $A_{element}$ is the area of the triangle plane. Similar equations can be formulated for the other unit displacements $d_{x2} = 1$, $d_{y2} = 1$ and $d_{x3} = 1$, $d_{y3} = 1$. The overall displacement of the planar element resulting from the several unit displacements is given by:

$$\left. \begin{array}{l} d_x(x,y) = f_1(x,y) d_{x1} + f_2(x,y) d_{x2} + f_3(x,y) d_{x3} \\ d_y(x,y) = f_1 d_{y1} + f_2 d_{y2} + f_3 d_{y3} \end{array} \right\} \quad V(x,y) = f \cdot V_k \quad (4.10)$$

with k = 1,2,3 for the displacement vector V_k. From Equation 4.10, the general form of the shear rate γ, and the strain ε (Equation 4.2, Equation 4.1) the following equation system can be deviated for element-wise constant shear stress and strain rate:

$$\varepsilon_x = \frac{\partial u}{\partial x} = \frac{1}{2A}\left((y_3 - y_2)\,d_{x1} + (y_1 - y_3)\,d_{x2} + (y_2 - y_1)d_{x3}\right)$$

$$= g_1 d_{x1} + g_2 d_{x2} + g_3 d_{x3}$$

$$\varepsilon_y = \frac{\partial v}{\partial y} = \frac{1}{2A}\left((x_2 - x_3)\,d_{y1} + (x_3 - x_1)\,d_{y2} + (x_1 - x_2)d_{y3}\right)$$

$$= g_4 d_{y1} + g_5 d_{y2} + g_6 d_{y3}$$

$$\gamma_{xy} = \frac{\partial u}{\partial y} + \frac{\partial v}{\partial x} = g_4 d_{x1} + g_5 d_{x2} + g_6 d_{x3} + g_1 d_{y1} + g_2 d_{y2} + g_3 d_{y3}$$

where g_i are coefficients of the shear stress and strain rate matrix, can be written as:

$$\varepsilon = G \cdot V_k \tag{4.11}$$

with G being the matrix for the coefficients g_i and V_k being the displacement vector. The general form of Hooke's law (Equation 4.4, Equation 4.5) can be used to formulate the relation between the shear stress and the strain rate according to the linear material behavior with temperature dependence. In accordance with Equation 4.11 the matrix form is set to be:

$$\sigma = E \cdot g \cdot V_k + \alpha_T \Delta T \tag{4.12}$$

The balance of forces for each element node for the displacements V_k can be expressed by the principle of virtual forces. The matrix formulation according to Bathe [6] is given by:

$$F_k \delta V_k^t = \iint_{(A)} \sigma\, \delta\, \varepsilon^t h\, dx\, dy \quad \rightarrow F = K_e \cdot V_k \tag{4.13}$$

where F_k is the vector of the nodal forces for a element, δ is the virtual displacement, t is the transpose of the matrix, and h the element thickness. Applying Equations 4.11 and 4.12 yields the expression on the right side of Equation 4.13, where K_e is the stiffness matrix of the element. For the whole structure with all elements the equilibrium equations for each element node has to be solved. This can be done in a direct fashion (Gauss algorithm, see Müller and Groth [58]) by superposing the individual stiffness matrices influencing the analyzed element node or mathematically, by applying a boolean matrix (iterative methods like Jacobi, Gauss-Seidel and Conjugate gradients [58]). For the external vector work force F_a it can be stated:

$$F_a = K \cdot V \tag{4.14}$$

which is a matrix system for all investigated element nodes with K being the system stiffness matrix. For the given displacement boundary conditions this system can be solved and the stresses at the nodes are derived using Equation 4.12. Several solver codes are provided within the FEM program ANSYS to execute these extensive calculations. To interpolate the displacements between the different nodes inside the element shape functions are applied. They can be linear or of higher order, dependent on the complexity of the observed structure and accuracy of the solution. Different element types using those functions are described in the following section.

4.3 Geometry and Grid Generation

The mesh for the geometry of the pressure vessel, its components and the steam plenum has to represent the geometrical features to an adequate resolution, so that the discretization error is small enough to satisfy the recommended accuracy. For the complex geometries, structured meshes are difficult to apply. Therefore unstructured meshes are preferred, since they are the most flexible and adaptive to an arbitrary solution domain boundary.

The disadvantage lies in the irregularity of the element and node structure which slows down the solution process compared to the structured meshes. The geometry of the reactor pressure vessel, its components and the steam plenum is generated with the Computer-Aided Design (CAD) package of the Computer Aided Three dimensional Interactive Application (CATIA) by Dassault. To ease the computational effort, the symmetric arrangement of the reactor pressure vessel flanges is used. Only a segment of the three dimensional geometry for the pressure vessel, closure head, the outlet pipes, the outlet flange with extension and closure head, and the connection tube to the turbine from the CAD-design is imported and defined as the computational domain. To represent the occurring loads, stresses and deformations, it is sufficient to model a segment of one-eighth or 45° of the axi-symmetric geometry.

The resulting computational domain can be seen in Figure 4.4. The same approach as for the reactor pressure vessel is used for the steam plenum. To represent the occurring loads, stresses and deformations it is sufficient to model a segment of one-eighth or 45° of the axis-symmetric geometry with one segment plane cutting through the outlet of the steam plenum and the other one through the middle of the guide strip. The resulting computational domain can be seen in Figure 4.5.

Different types of elements can be used to discretize the three dimensional numerical domain. For the quite complex geometry of the reactor pres-

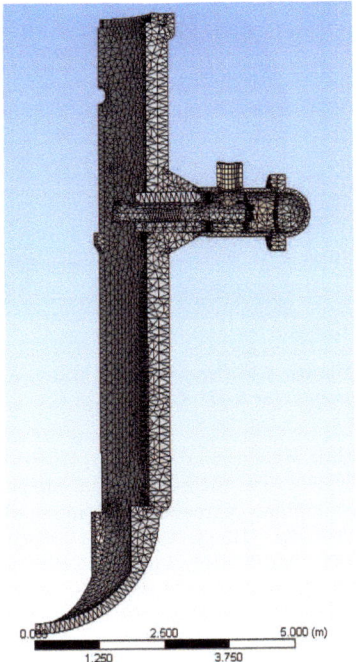

Figure 4.4: Computational grid in ANSYS from the imported CATIA geometry of the reactor pressure vessel and components for a 45° segment.

sure vessel, its components and the steam plenum, the implemented automatic mesh generator ANSYS WORKBENCH is used to generate the computational mesh. Two different three dimensional solid element types have been chosen for the mesh generation together with two other types which are responsible for the connection of the different components. They allow for a good representation of the geometry in due respect of the thermo-mechanical problem and are suitable for the problem-related and kinematic boundary conditions, like load application. SOLID187 is a three dimensional 10-node tetrahedral structural solid, which has a quadratic displacement behavior and is suitable to model irregular meshes. Additionally to the four corner nodes of the four triangle planes it features 6 nodes bisecting the edge lengths of the triangle planes. Elements with midside nodes feature a high quality, they are capable of forming curved edge shapes resembling the borders of the investigated geometry. The other element is the

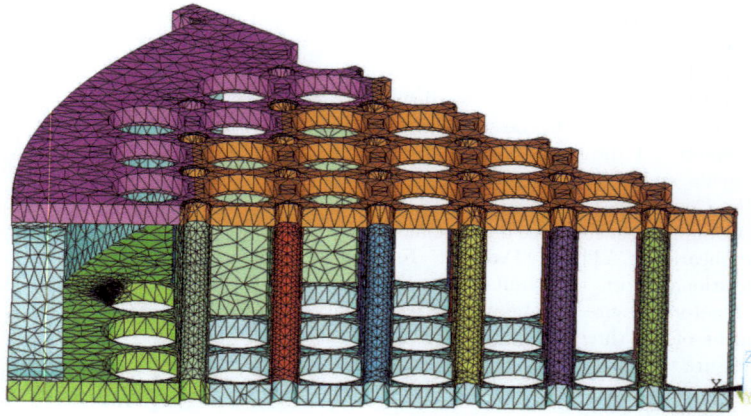

Figure 4.5: Computational grid in ANSYS from the imported CATIA geometry of the steam plenum for a 45° segment.

three-dimensional 20-node structural element SOLID186, which is also of higher order and occurs in three different shapes: tetrahedral, pyramid, and prism. Both elements have three degrees of freedom per node and can be adapted precisely to the shape of the geometry. Due to the higher order approximation scheme with a quadratic displacement behavior, a lower approximation error compared to less complex elements is observed (Bathe [6]).

The two elements CONTA170 and CONTA174 are used to represent contact and sliding between different components inside the computational domain. They can either have the same geometric characteristics as the solid element face (CONTA174) or the target surface of the solid element is modeled through a set of target segments comprising the target surface (CONTA170).The quality of the meshes for the different components has been checked in relation to the connectivity of all external nodes of neighboring elements, aspect ratio of the edges below 1:10, and allowed included angle of cell faces. Additionally, the transition from large to small elements is executed in a gradual manner to minimize

4.4 Coupled Thermal-Structural Analysis

numerical diffusion effects for the equation matrix. The junction nodes are selected in such a way, that the calculation result is sufficiently exact for the problem to be solved. For strongly varying variables (stresses or strain), the fineness of the mesh is adapted accordingly, performing a grid sensitivity study. One example for such a refinement can be seen in Figure 4.6 for the evaluation of the von Mises stress distribution in the vicinity above the flange inlet. Starting with a coarse mesh (A), several refinement steps

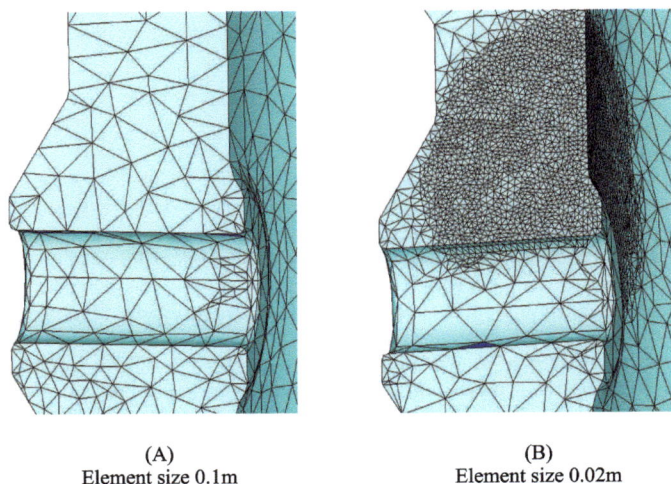

(A) (B)
Element size 0.1m   Element size 0.02m

Figure 4.6: Generated mesh for the inlet flange (A) with the locally refined region in the vicinity of the upper flange (B) for the grid sensitivity study.

have been applied until the resulting stress distribution deviation between two successive steps stays inside a prescribed limit of 1 MPa. The comparison of the different von-Mises stress distributions shows, that the refinement with an average element size of 0.02 m (B) exhibits a sufficiently small strees intensity deviation compared to a coarser mesh with 0.05 m.

4.4 Coupled Thermal-Structural Analysis

The combined thermo-mechanical analysis is performed in a sequentially coupled physics analysis in ANSYS, where the structural analysis depends on the results of the thermal analysis. Due to the different physical conditions each environment has to be created separately, but a single set of nodes can be used for the entire model. The geometry is created for the thermal analysis and kept constant for the coupled analysis. To model the different physical effects, the element type has to be changed from thermal elements to structural elements between the analyses. Only certain combinations of elements are allowed in ANSYS, for the mentioned structural elements SOLID187 and SOLID186 the corresponding counterparts are named SOLID87 and SOLID90. The first element, SOLID87 is a three dimensional 10-node tetrahedral thermal solid, which is well suited to model irregular meshes. The second one, SOLID90, is also a three dimensional but 20-node thermal solid, which is able to model curved boundaries. Both elements have one single degree of freedom, temperature, at each node. The thermal properties of the material are set with the conductivity, element loads for both solids involve convection, temperature or heat flux. The result of the thermal analysis serves as a boundary condition for the structural analysis, for which the element types and the material properties are modified to satisfy the new physical environment. Subsequentely, the structural analysis with the individual constraints and the thermal effects from the first analysis is carried out.

4.5 Boundary Conditions

To solve the stiffness matrix system of the investigated problem, constraints have to be applied as initial condition. Those constraints can be derived from the boundary conditions of the investigated structure. A particle that moves in three dimensional space has three translational displacement components as Degrees of Freedom (DOF), while a rigid body would have at most six DOFs including three rotations. For the finite element approach each element node has three translational displacement components as DOF, which are used to prevent free movement of the investigated components in space. The restriction of the structure in a certain direction corresponds to the suppression of one DOF. To determine the stresses and displacements, the system has to be kinematically stable, so that no rigid body movements are allowed. Restrictions for the DOF can be applied directly on the element nodes or on the element surface.

Due to the symmetry boundary condition, the following restrictions are applied for all investigated components. For a cartesian coordinate system, the component is positioned in a way that expansions and deformations in radial direction are possible. Therefore, one of the symmetry planes is restricted for movement orthogonal to the plane. The second restriction limits movement in z-direction, to prevent solid body movement.

Constraints also involve all kinds of loads, which act on the structure. In ANSYS, initial state loadings are defined to be of two types: nodal and element. Nodal loads such as nodal force loads are associated with the DOF at the node. Element loads are surface loads, body loads, and inertia loads. They are always associated with a particular element. Surface loads have the input pressure (or force per area) for structural elements or convection for thermal elements. Body loads are only applied as temperatures for structural elements. Gravity is the only inertial load that is applied with the structural elements having mass. The occurring loads are applied to the several components of the coupled thermal-structural analysis, a detailed description of all applied loads is given in chapter 11, page 91.

5 Computational Fluid Dynamics

The application of computational fluid dynamics (CFD) for the fluidic optimization of components has several advantages compared to experiments at the early stage of the design process. Due to the application of the software on personal computers, the simulations are less costly and time consuming than experiments. It is possible to analyze different flow phenomena with one calculation, which makes this tool very effective for optimization processes, where each improvement step results in a geometrical modification for the apparatus and investigation tools in the experimental approach. Additionally, boundary conditions are easily prescribed for CFD-applications compared to the experimental effort.

Computational fluid dynamics use detailed solutions of the Navier-Stokes equations as substitutes for experimental research of complex three dimensional flows and geometries, where the simple flow analysis approach fails. In many cases turbulence plays an important role and has to be modeled mathematically. Since the exact equations are not available and feasible, the introduction of useful models to reduce costs and computational time is necessary. Those models rely on the simplification of the exact conservation laws, but have to be customized for the application, since there is no general solution method.

The finite volume method is chosen for the discretization of the differential equations using a set of algebraic equations. This approach is very robust and can be formulated directly in physical space. The results have to be discussed in relation to approximations made in the discretization process to optimize the accuracy of the method. A good compromise has to be found, since the use of more accurate interpolations and approximations to smaller regions is more time consuming and costly. The solving of the discretized equations is done with iterative methods as direct solvers are normally too costly. Iterative methods, on the other hand could cause errors due to stopping the iteration process too soon, in order to save computational time. Therefore, an error estimation and analysis is very important and imminent for the obtained solution.

The discrete locations, at which the variables are calculated, are defined by the numerical grid, in the case of the finite volume approach the solution domain is divided in control volumes. For simple geometries the numerical grid can be structured, where any control volume within the calculation domain is uniquely identified by a set of three indices for a 3D geometry. Unstructured grids are used for very complex geometries, since they are the most flexible type of grid. The disadvantage lies in the irregularity of the data structure, which causes the solver to be much slower than for the regular grids.

5.1 Basic Conservation Equations

The definition of the general fluid equations of motion as described in Ferziger and Perić [23] are used to derive the appropriate formulation for the equations to describe the flow within the backflow limiter. Several assumptions are made for the calculations. First, the fluid water is defined as incompressible, since no two-phase flow is allowed for the calculations. Second, the fluid is regarded as Newtonian with the viscosity being constant throughout the flow. Therefore, the continuity equation for a three dimensional, incompressible, and stationary flow can be written as follows, according to Oertel and Laurien [63]

$$\frac{\partial u}{\partial x} + \frac{\partial v}{\partial y} + \frac{\partial w}{\partial z} = 0 \tag{5.1}$$

which can be rewritten in Eulerian vector form

$$\nabla \cdot U = 0 \tag{5.2}$$

where the Nabla operator is defined as $\nabla = \left(\frac{\partial}{\partial x}, \frac{\partial}{\partial y}, \frac{\partial}{\partial z}\right)^T$. The momentum equations for a cartesian coordinate system with x, y and z direction take the following form

$$\rho \cdot \left(u \cdot \frac{\partial u}{\partial x} + v \cdot \frac{\partial u}{\partial y} + w \cdot \frac{\partial u}{\partial z}\right) = F_x - \frac{\partial p}{\partial x} + \mu \cdot \left(\frac{\partial^2 u}{\partial x^2} + \frac{\partial^2 u}{\partial y^2} + \frac{\partial^2 u}{\partial z^2}\right) \tag{5.3}$$

$$\rho \cdot \left(u \cdot \frac{\partial v}{\partial x} + v \cdot \frac{\partial v}{\partial y} + w \cdot \frac{\partial v}{\partial z}\right) = F_y - \frac{\partial p}{\partial y} + \mu \cdot \left(\frac{\partial^2 v}{\partial x^2} + \frac{\partial^2 v}{\partial y^2} + \frac{\partial^2 v}{\partial z^2}\right) \tag{5.4}$$

$$\rho \cdot \left(u \cdot \frac{\partial w}{\partial x} + v \cdot \frac{\partial w}{\partial y} + w \cdot \frac{\partial w}{\partial z}\right) = F_z - \frac{\partial p}{\partial z} + \mu \cdot \left(\frac{\partial^2 w}{\partial x^2} + \frac{\partial^2 w}{\partial y^2} + \frac{\partial^2 w}{\partial z^2}\right) \tag{5.5}$$

These equations are also called Navier-Stokes equations and can be rewritten in Eulerian vector form

$$\rho \cdot ((U \cdot \nabla) \cdot U) = F - \nabla p + \mu \cdot \Delta U \tag{5.6}$$

with the pressure gradient ∇p,

$$\nabla p = \left(\frac{\partial p}{\partial x}, \frac{\partial p}{\partial y}, \frac{\partial p}{\partial z}\right)^t \tag{5.7}$$

the convection operator $(U \cdot \nabla) \cdot U$ with the scalar product of the velocity vector U and the Nabla operator $U \cdot \nabla$,

$$U \cdot \nabla = u \cdot \frac{\partial}{\partial x} + v \cdot \frac{\partial}{\partial y} + w \cdot \frac{\partial}{\partial z} \tag{5.8}$$

and ΔU as the Laplace operator applied to the velocity vector U.

$$\Delta U = \frac{\partial^2 U}{\partial x^2} + \frac{\partial^2 U}{\partial y^2} + \frac{\partial^2 U}{\partial z^2} \tag{5.9}$$

These continuity and momentum equations are decoupled from the energy equation and are all that is necessary to solve for the velocity and pressure fields in incompressible, laminar flow. However, for the application in the numerical flow analysis of the backflow limiter with Reynolds numbers above 10^5, these equations have to be extended to turbulent flows.

5.2 Turbulence Modeling

For turbulent flows, the Reynolds averaging forms the basis of the applied Reynolds-averaged Navier Stokes Equations. As an example, the velocity component u is represented by

$$u = \bar{u} + u' \tag{5.10}$$

where u is the sum of u' which is defined as the fluctuation or perturbation term for the velocity u such that $\bar{U}' = 0$. $\bar{u}$, which is the time averaged mean velocity in x-direction with the definition

$$\bar{u} = \frac{1}{t_2 - t_1} \cdot \int_{t_1}^{t_2} u(x, y, z, t) \cdot dt \tag{5.11}$$

where the period $\delta t = t_2 - t_1$ is chosen to be sufficiently large in order to smooth out u but sufficiently small with respect to the transient time constants. For an incompressible fluid, the continuation equation can be written in the tensor form as follows

$$\frac{\partial \bar{u}_i}{\partial x_i} = 0 \tag{5.12}$$

$$\underbrace{\rho \cdot \left(\frac{\partial}{\partial x_j} (\bar{u}_j \cdot \bar{u}_i) \right)}_{\text{Convection term}} = \frac{\partial \bar{p}}{\partial x_i} + \underbrace{\frac{\partial}{\partial x_j} \cdot \left(\mu \cdot \left(\frac{\partial \bar{u}_i}{\partial x_j} + \frac{\partial \bar{u}_j}{\partial x_i} \right) - \rho \cdot \left(\overline{u'_i \cdot u'_j} \right) \right)}_{\text{Diffusion term} \quad \text{Reynolds stresses}} + \underbrace{F_i}_{\text{Source term}} \tag{5.13}$$

The presence of the Reynolds stresses $-\rho \cdot \left(\overline{u'_i \cdot u'_j} \right) = \tau^t_{ij}$ in the conservation equations means that the latter are not closed, since they contain more variables than there are equations. The nine Reynolds stresses from the 3×3 matrix can be reduced due to the symmetry of the matrix to six variables. Closure requires use of some approximations, which usually take the form of prescribing the Reynolds stress tensor in terms of the mean quantities. In the following, approximations are introduced, which are called turbulence models in engineering.

5.2.1 Eddy Viscosity Models

One possible approach is that the effect of turbulence can be represented as an increasing viscosity, which is the normal mechanism for laminar flows, where the energy dissipation and transport of mass, momentum and energy normal to the streamlines is mediated by the viscosity. This leads to the eddy viscosity models for the Reynolds stress using the Boussinesq approach, where μ_t is the turbulent viscosity [63]

$$-\rho \cdot \left(\overline{u'_i \cdot u'_j} \right) = \mu_t \cdot \left(\frac{\partial \bar{u}_i}{\partial x_j} + \frac{\partial \bar{u}_j}{\partial x_i} \right) - \frac{2}{3} \cdot \rho \cdot \delta_{ij} \cdot k \tag{5.14}$$

k is the turbulent kinetic energy and defined as

$$k = \frac{1}{2} \overline{u'_i \cdot u'_j} = \frac{1}{2} \left(\overline{u'_x u'_x} + \overline{u'_y u'_y} + \overline{u'_z u'_z} \right) \tag{5.15}$$

The last term on the right side of Equation 5.14 represents the turbulent pressure and acts as a correction term to close the equation. In the simplest description for the eddy viscosity models, turbulence can be characterized by two parameters, its kinetic energy k, and a length scale L. Here, k is determined from the mean velocity field and L is a prescribed function of the given coordinates. This approach is only possible for simple flows, but not for highly complex three-dimensional flows. Therefore, more complex models are introduced, where the dissipation ε is linked to the kinetic energy k and the length scale L. The characterization of the turbulence with two transport values k and ε is based on the idea that the creation and dissipation of turbulent structures can be seen as an energy cascade. For high Reynolds numbers there is a cascade of energy from the largest scales to the smallest ones and the energy transfered to the smallest scales is dissipated. The k-ε model uses this approach, where the large scale eddies are associated with the turbulent kinetic energy k and the small scale eddies are connected to the dissipation rate ε. The transport equation for the **turbulence kinetic energy** takes the following form (Wilcox [96])

$$\underbrace{\frac{\partial(\bar{\rho}\bar{u}_j k)}{\partial x_j}}_{\text{Convective transport}} = \underbrace{\frac{\partial}{\partial x_j}\left(\mu \frac{\partial k}{\partial x_j}\right)}_{\text{Molecular diffusion}} - \underbrace{\frac{\partial}{\partial x_j}\left(\frac{\rho}{2}\overline{u'_j u'_i u'_i} + \overline{p' u'_j}\right)}_{\text{Turbulent diffusion}} - \underbrace{\rho\overline{u'_i u'_j}\frac{\partial \bar{u}_i}{\partial x_j}}_{P_k} - \underbrace{\mu\overline{\frac{\partial u'_i}{\partial x_k}\frac{\partial u'_i}{\partial x_k}}}_{\text{Dissipation}} \quad (5.16)$$

The turbulent diffusion term of kinetic energy can be approximated by

$$-\left(\frac{\rho}{2}\overline{u'_j u'_i u'_i} + \overline{p' u'_j}\right) \approx \frac{\mu_t}{\text{Pr}_t}\frac{\partial k}{\partial x_j} \quad (5.17)$$

where μ_t is the dynamic eddy viscosity defined above and Pr_t is the turbulent Prandtl number for k whose value is approximately unity. The rate of production of turbulent kinetic energy by the mean flow, can be approximated using Equation 5.14 as

$$P_k = -\rho\overline{u'_i u'_j}\frac{\partial \bar{u}_i}{\partial x_j} \approx \mu_t \cdot \left(\frac{\partial \bar{u}_i}{\partial x_j} + \frac{\partial \bar{u}_j}{\partial x_i}\right)\frac{\partial \bar{u}_i}{\partial x_j} \quad (5.18)$$

The **turbulence dissipation rate** can be written in its most common form as

$$\frac{\partial(\rho u_j \varepsilon)}{\partial x_j} = C_{\varepsilon 1} \cdot P_k \cdot \frac{\varepsilon}{k} - \rho \cdot C_{\varepsilon 2} \cdot \frac{\varepsilon^2}{k} + \frac{\partial}{\partial x_j}\left(\frac{\mu_t}{\text{Pr}_{t\varepsilon}}\frac{\partial \varepsilon}{\partial x_j}\right) \quad (5.19)$$

In the standard k - ϵ turbulence model, the eddy viscosity is expressed as

$$\mu_t = \rho \cdot C_\mu \cdot \sqrt{k} \cdot L = \rho \cdot C_\mu \cdot \frac{k^2}{\varepsilon} \quad (5.20)$$

$\text{Pr}_{t\varepsilon}$ is the turbulent Prandtl number for ε whose value is approximately 1.22, $C_{\varepsilon 1}$, $C_{\varepsilon 2}$ are coefficients. The disadvantage of the k-ε model are the overprediction of turbulence for flows with pressure gradients and stagnation points which results in the underprediction of recirculation areas. Therefore, a second two equation eddy viscosity model is introduced, the k-ω turbulence model. It uses the turbulent frequency ω instead of the turbulent dissipation ε to describe turbulent flows.

$$\omega = \frac{1}{C_\mu}\frac{k}{\varepsilon} \quad (5.21)$$

5.2 Turbulence Modeling

The two transport equations are similar to the k-ε model approach and can be written according to Wilcox [96]:

Turbulence kinetic energy

$$\frac{\partial (\rho u_j k)}{\partial x_j} = P_k - \beta^* \rho k \omega + \frac{\partial}{\partial x_j}\left(\left(\mu + \frac{\mu_t}{\text{Pr}_{tk}}\right)\frac{\partial k}{\partial x_j}\right) \quad (5.22)$$

Specific dissipation rate

$$\frac{\partial (\rho u_j \omega)}{\partial x_j} = \alpha \frac{\omega}{k} P_k - \beta \rho \omega^2 + \frac{\partial}{\partial x_j}\left(\left(\mu + \frac{\mu_t}{\text{Pr}_{t\omega}}\right)\frac{\partial \omega}{\partial x_j}\right) \quad (5.23)$$

where the eddy viscosity μ_t is defined as

$$\mu_t = \rho \frac{k}{\omega} \quad (5.24)$$

β^*, α, β are closure coefficients, and Pr_{tk}, $\text{Pr}_{t\omega}$ is the turbulent Prandtl number for k whose value is 2.0 and ω whose value is 2.0. The k-ω model has the disadvantage of a strong sensitivity to the values of ω in the free stream outside the boundary. A model is required that combines the advantages of both models.

5.2.2 k-ω SST model

The k-ω Shear Stress Transport (SST) model by Menter [57] uses a zonal approach where a blending of k-ω at the wall and k-ε in the free stream is used. Additionally, the empirical constant C_μ is modified to predict more precisely flows with adverse pressure gradients and separation. The **turbulence kinetic energy** equation is the same as for the k-ω approach, only the definition of the eddy viscosity is modified to account for the transport of the principal turbulent shear stress. For the implementation in the model the coefficients have to be adapted:

$$\frac{\partial (\rho u_j k)}{\partial x_j} = P_k - \beta^* \rho k \omega + \frac{\partial}{\partial x_j}\left(\left(\mu + \frac{\mu_t}{\tilde{\text{Pr}}_{tk}}\right)\frac{\partial k}{\partial x_j}\right) \quad (5.25)$$

and for the **specific dissipation rate**

$$\frac{\partial (\rho u_j \omega)}{\partial x_j} = \tilde{\alpha}\frac{\omega}{k} P_k - \tilde{\beta} \rho \omega^2 + (1-F_1)\frac{2}{\text{Pr}_{t\omega}}\frac{\partial k}{\partial x_j}\frac{\partial \omega}{\partial x_j} + \frac{\partial}{\partial x_j}\left(\left(\mu + \frac{\mu_t}{\tilde{\text{Pr}}_{t\omega}}\right)\frac{\partial \omega}{\partial x_j}\right) \quad (5.26)$$

where the coefficients $\tilde{\alpha}, \tilde{\beta}, \tilde{\text{Pr}}_{t\omega}$ are a function of F_1 in the kind of

$$\tilde{\alpha} = \alpha_{k-\omega} \times F_1 + \alpha_{k-\varepsilon}(1-F_1) \quad (5.27)$$

The blending function F_1 is defined by

$$F_1 = \tanh\left(\left(\min\left(\max\left(\frac{\sqrt{k}}{\beta^* \omega y}, \frac{500\nu}{y^2 \omega}\right), \frac{4\rho \tilde{\text{Pr}}_{t\omega} k}{CD_{k\omega} y^2}\right)\right)^4\right) \quad (5.28)$$

where $CD_{k\omega} = \max\left(2 \cdot \rho \cdot \tilde{\text{Pr}}_{t\omega} \cdot \frac{1}{\omega} \cdot \frac{\partial k}{\partial x_j} \cdot \frac{\partial \omega}{\partial x_j}, 10^{-20}\right)$ and y is the distance to the nearest wall. The blending function F_1 is designed to be one in the sublayer and logarithmic region of the boundary layer representing the k-ω model and to gradually switch to zero in the wake region representing the k-ε model. While the standard k-ε and k-ω models overpredict the shear stresses for flows with pressure gradients and therefore result in underprediction or neglect of separation, the SST model forces the Bradshaw relation giving a better prediction of separation:

$$-\overline{uv} = a_1 \cdot k \qquad (5.29)$$

where a_1 represents a coefficient with the value of 0.31. The turbulent eddy viscosity ν_t is described as

$$\mu_t = \rho \frac{a_1 k}{\max(a_1 \omega, \Omega F_2)} \qquad (5.30)$$

here, Ω is the absolute value for the vorticity and equals $\frac{\partial u}{\partial y}$ and F_2 is a second blending function defined by

$$F_2 = \tanh\left(\left(\max\left(\frac{2\sqrt{k}}{\beta^* \omega y}, \frac{500 \nu}{y^2 \omega}\right)\right)^2\right) \qquad (5.31)$$

which is one for boundary layer flows and zero for free shear layers. A production limiter is then used in the SST model to prevent the build-up of turbulence in stagnation regions

$$P_k = \mu_t \cdot \frac{\partial u_i}{\partial x_j}\left(\frac{\partial u_i}{\partial x_j} + \frac{\partial u_j}{\partial x_i}\right) \quad \rightarrow \quad \tilde{P}_k = \min(P_k, 10 \cdot \beta^* \rho k \omega) \qquad (5.32)$$

5.2.3 Non-Linear Models

The two-equation turbulence models presume that the turbulent stresses are linearly related to the rate of strain by a scalar turbulent viscosity, meaning that the turbulence in a three dimensional flow is transported as a scalar which is the same for all three directions. The assumption of isotropic turbulence is only valid for fairly simple states of strain, where the coefficients have been validated by similar cases. For flows in complex geometries or flows with swirl, this assumption does not hold true anymore and an anisotropic turbulence model has to be used. One possible approach is to determine the turbulent stresses directly by solving a transport equation for each stress component. These so called Reynolds stress transport models (RSM) are complex and expensive to compute, leading to a minor use in industrial applications. Another, simpler approach are non-linear eddy viscosity models, which still relate the turbulent stresses to the rate of strain, but higher order quadratic and cubic terms are included compared to the linear eddy viscosity models. They are computationally efficient as they involve the same number of equations as two-equation models.

5.3 Discretization Method

For the spatial discretization of the flow, the finite volume method as described in Ferziger [23], Oertel and Laurien [63], and Versteeg and Malalasekera [92] is used. The solution domain is subdivided into a finite number of small control volumes by a grid, which

5.3 Discretization Method

defines the control volume boundaries. A usual approach defines the control volumes by a suitable grid and assigns the computational node to the control volume center. Here, the differential equations governing mass and momentum within the fluid are discretized directly in physical space compared to the piecewise polynomial functions on local elements for the finite elements method. The net flux through the control volume for the nodal center is defined as the sum of integrals over all six control volume surfaces. In order to calculate the surface integral exactly, one would need to know the integrand everywhere on the surface. However, this is not possible since only the values at the nodal center are calculated. Therefore, the cell-face values are approximated in terms of the nodal control volume center value. The midpoint rule is the simplest approximation for the surface integral with second order accuracy, which in this case is defined as the product of the mean value over the surface at the surface center, also called the integrand and the surface area. Since the integrand is not available at the surface center, it has to be obtained by interpolation. In order to preserve the second order accuracy of the midpoint rule approximation, these interpolations also have to have at least the same accuracy. The applied interpolation methods are presented in the following subsection. The terms of the integrated equations which are substituted by the difference-type approximations for the cell-centered nodal values are then solved by those iterative methods.

One advantage of the finite volume method is the use of unstructured meshes, since the method works with the cell volumes and not the grid intersection points. For complicated geometries, this option offers a greater flexibility for the adaptation of custom meshes which cannot be identified with coordinates lines. The commercial CFD-code STAR-CD offers a variety of cell shapes and combinations of mesh structures to fill any complex flow volume using the finite volume approach.

5.3.1 Interpolation Methods

The approximations of the integrals at the surface require the values of variables at locations other than the center of the control volume. They have to be expressed in terms of the nodal values at the center by interpolation. Numerous possibilities are available and can be found in the literature (Ferziger [23], Oertel and Laurien [63]), the ones presented here are applied and compared for the executed numerical analysis.

Upwind Differencing Scheme (UD)

This low first-order scheme by Courant et al. [19] selects the velocities of the nearest upwind or upstream node value to approximate the considered convective flux through the cell face downstream of that node. This is equivalent to using a backward- or forward-difference approximation for the first derivation, depending on the flow direction. In other words, the node value of a transport scalar upstream is used for the discretization, which can also be explained using the physical background where the momentum is coupled to the mass which moves with a certain velocity downstream. The value of a certain cell volume depends mainly on the value of the following cell upstream. With this method a first order error reduction is obtained, which results in a robust non oscillating solution, but achieves this by being numerically diffusive since the midterm rule is second order accurate. For flows with peaks or rapid variations of the variables a very fine grid is necessary to obtain accurate solutions. Therefore the UD is used to yield a first convergent solution and then switch to a more accurate numerical scheme like the QUICK interpolation to improve the convergence and accuracy of the solution.

Quadratic Upwind Interpolation of Convective Kinematics (QUICK)

This third order scheme by Leonard [53] approximates the variable profile between two following control volumes by a parabola through two points upstream and one point downstream to get an interpolated value. Therefore, it extends the computational range one more point in each direction. Its quadratic formulation has a third order accuracy (Leonard [54]) for both structured and unstructured meshes. It should not be used for tetrahedral meshes and can cause non-physical spatial oscillations, which result in numerical dispersion for extremely high gradients of one variable. Those problems can be omitted by using a blending of the QUICK approach and the UD approach which is used as a damping function. For the backflow limiter, this blending is applied to properly solve the pressure variables as they exhibit high gradients throughout the flow.

5.4 Numerical Solution Method

The result of the discretization practice is a system of algebraic equations, which have a non-linear character according to the nature of the partial differential equations from which they are derived. The CFD-code STAR-CD employs several implicit methods to solve those algebraic equations, in the case of steady-state calculations the applied implicit approach is the Semi-Implicit Method for Pressure-Linked Equations (SIMPLE) algorithm which has been introduced by Patankar and Spalding [69]. Due to the non-linearity and coupling of the underlying differential equations they cannot be solved directly, but iteratively. In general the approach for the iterative method is the same as for the unsteady solution, the steady problem is regarded as solving an unsteady problem until a steady solution is reached. The main difference lies in the size of the time step, compared to unsteady calculations it is chosen to be large enough to reach the steady state quickly and not to obtain an accurate time history as for the unsteady solution.

To solve the non-linear equations, a so called predictor-corrector strategy is applied, where first an initial pressure distribution is estimated and a provisional velocity field is derived from the momentum equations by the solution sequence. Continuity is enforced for this velocity field by applying a so-called pressure correction equation, which is in turn used to correct the pressure field. Then the initial velocity field is refined in several corrector stages using under-relaxation as a stabilizing measure until both momentum and continuity balances are satisfying the demanded tolerance. To solve the decoupled linearized algebraic equations for each variable in the momentum equations two techniques are applied. The conjugate gradient (CG)-type solver with various preconditioning methods and the algebraic multigrid (AMG) approach, which uses multigrid methods of solving the matrix equations without relying on the geometry of the problem being solved. The CG-solver is used for structured meshes with low geometrical complexity, where it gives good convergence behavior. In the case of a numerical mesh with unstructured and very complex grids, the CG-solver is not applicable, since the convergence behavior of the CG-solver is sensitive to meshes with varying cell sizes and causes problems for the convergence of the pressure correction equation. Therefore, the multigrid technique of the AMG-solver is used, where the errors of each step for the pressure correction equation are smoothed out on a grid that is most suitable for this purpose. The method cycles between coarser and finer grids to solve the equations using transfer of residuals from fine to coarse grids and interpolation of corrections from coarse to fine meshes. This approach is a very efficient convergence method for fine grids and is also very economical for coarse grids.

5.5 Geometry and Grid Generation

The numerical grid has to fit the boundary surface of the computational domain and be able to subdivide its volume into a finite number of control volumes or cells for the spatial discretization. Furthermore, it has to represent the geometry and geometrical features to an adequate resolution so that the discretization error is small enough to satisfy the recommended accuracy. Several kinds of mesh topology are available to represent the geometrical features.

The regular or structured grid is the simplest mesh structure and consists of hexahedral control volumes. The position of any hexahedron within the numerical domain is uniquely defined by a set of three indices representing the cartesian coordinates x, y and z. Each point has exactly six neighbors, which simplifies programming and reduces computational effort, since the algebraic equation system has a regular structure. The disadvantage of structured grids lies in their limited flexibility and higher time and effort to generate a mesh for complex geometries. A second type of grid, the so-called block-structured grid, uses hexahedrons and therefore has a better geometrical flexibility. It divides the computational domain in several suitable blocks, which are then defined as a structured grid on the fine level. Then the different block interfaces are matched using a special coupling technique. This kind of grid allows block-wise local refinement to adapt to flow regions where better resolution is required. In the case of complex geometries, where a high number of blocks is required to represent the computational domain, this approach is complicated and inefficient.

For complex geometries, unstructured grids can be applied, since they are the most flexible and adaptive to an arbitrary solution domain boundary. They consist of straight-edged cells of various forms (hexahedron, tetrahedron, prism, pyramid and also other polyhedral shapes) and may be used in an arbitrary manner to fill any volume. They can be generated automatically with easy control of the aspect ratio and local refinement. The disadvantage lies in the irregularity of the data structure, which slows down the solution process for the algebraic equations compared to the structured grids.

The mesh generator tool pro-am of the CFD-code STAR-CD uses a combination of different polyhedral cells to model the numerical domain. To enhance the grid quality in the area of boundaries such as inlet, outlet, and wall a layered prism structure is used, which is called extrusion layer or subsurface. This layer has been applied for all wall surfaces in order to have a high resolution for the transition between boundary layer and free flow. To adapt the grid in critical regions with high flow gradients, different embedded refinements have been chosen, where the initial grid is subdivided locally in order to increase the resolution. Figure 5.1 shows a numerical grid for the backflow limiter consisting of approximately 135 400 trimmed cells for the free flow zone and 140 500 prism cells for the extrusion layer.

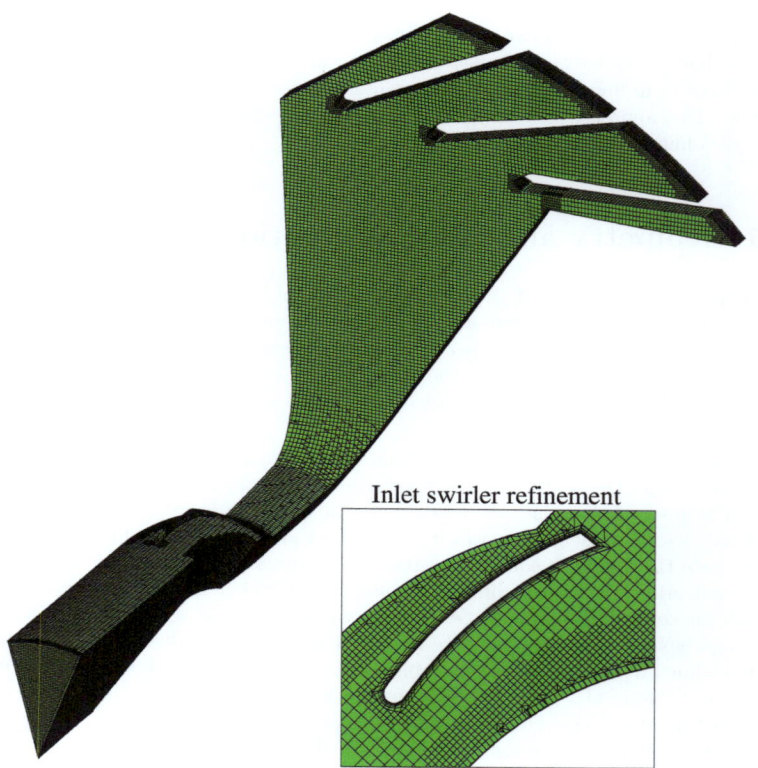

Figure 5.1: Generated hybrid mesh of the backflow limiter segment with one inlet swirler vane (detailed refinement on right side) and three swirler vanes.

As an example the grid refinement around the inlet swirler vane for one cutting plane in horizontal direction is displayed in Figure 5.1 in the lower right corner. The refinement toward the wall has been adapted to satisfy the y^+ region which is required to satisfy the laws-of-the-wall for the applied turbulence models. The quality of the mesh has been improved in relation to aspect ratio, skewness, and allowed included angle of adjacent faces.

5.6 Boundary Conditions

The Navier Stokes equations represent a system of differential equations which can only be solved for prescribed initial and boundary conditions. A constant distribution of all flow variables throughout the flow domain is normally chosen as initial condition, since there is no information available for the real flow distribution at the beginning. For cases where even this information is not known, stagnant conditions can be assumed

5.6 Boundary Conditions

where the velocity throughout the domain is chosen as zero in every cell volume (zero-flow condition). Values for the turbulent quantities k, ε, or ω are extrapolated from the average values of the inlet flow conditions.

The boundary conditions allow the coupling of the fluid domain with the surroundings that are not part of the simulation. The finite volume method requires that the boundary fluxes are either be known or expressed in terms of known quantities. Two types of boundary conditions and combinations of them are generally encountered for this method. The Dirichlet condition specifies the distribution of a physical quantity over the boundary and can be used to prescribe for example a mass flow at the inlet of the computational domain or the no-slip condition at the wall. The Neumann condition defines the distribution of the first derivative of a physical quantity normal to the boundary layer. Outlet boundaries are normally described by the gradient of the variables of interest in flow direction. It is also possible to have linear combinations of both types.

Inlet boundary The flow variables velocity for all three directions and the density have to be prescribed at the boundary to solve the Navier-Stokes equations, while the pressure is extrapolated from the flow domain. The distribution of those flow variables needs to be specified for the cross section of the inlet boundary using either the Dirichlet or the Neumann condition.

Outlet boundary The four variables and the turbulent quantities k, ε, or ω are also needed at this type of boundary in analogy to the inlet boundary. For the outlet boundary, they are extrapolated from the flow domain. The boundary condition imposed at the outlet should have a weak influence on the upstream flow. The location of the outlet has to be selected far away from geometrical disturbances, so that the flow reaches a fully developed state where no change occurs in the flow direction. Then the Neumann condition is satisfied and the gradients of all variables except the pressure are zero in the outward direction.

Wall boundary for turbulent flow The flow throughout the backflow limiter and in the vicinity of the walls is considered to be turbulent. In wall attached boundary layers, the normal gradients in the flow variables become extremely large as wall distance reduces to zero, since the viscous fluid sticks to solid boundary. A large number of cell layers close to the wall is required to resolve these gradients increasing with associated computing overheads. Additionally, as the wall is approached, viscous effects become dominant compared to the turbulent fluctuations and the standard turbulence models are not valid all the way through to the wall. Thus, special wall modeling procedures have to be used.

For the k-ε and the ω-SST model, the near-wall region is not explicitly resolved but is bridged using so called wall functions, which are explained in Rodi [72] and Wilcox [96]. In order to characterize the wall functions, the region close to the wall is expressed with variables written in a dimensionless manner with respect to conditions at the wall. Furthermore, variations in the variables are predominantly normal to the wall leading to a one-dimensional behavior with an assumed uniform shear stress distribution in the layer. Additionally, turbulence energy production and dissipation are balanced with a linear variation of turbulence length scale. The profile of the dimensionless velocity u^+

in terms of the normal distance y from the wall can be written as:

$$u^+ = \begin{cases} y^+, & y^+ \leq y_m^+ \\ \frac{1}{\kappa} \ln(Ey^+), & y^+ > y_m^+ \end{cases} \quad (5.33)$$

where u^+ is defined as $\frac{u}{u_\tau}$ and the dimensionless wall distance y^+ as $\frac{y \cdot \rho \cdot u_\tau}{\mu}$ with the wall friction velocity u_τ equaling $\sqrt{\frac{\tau_w}{\rho}}$ represents the wall shear stress, u is the time-averaged velocity parallel to the wall, E is an empirical coefficient and κ the von Karman constant. y_m^+ defines the limit between the viscous sublayer and the so-called log-layer and satisfies the condition $y_m^+ - \frac{1}{\kappa} \ln(Ey_m^+) = 0$. A linear relationship is observed between the velocity u^+ and y^+ in the viscous sublayer, and a logarithmic relationship the law of the wall in the adjacent layers. These functions are used to relate flow variables at the first computational mesh central node directly to the wall shear stress, without resolving the mesh in between. The grid has to be arranged in a way that the values of y^+ at the wall adjacent central node do not exceed 100 and are greater than 30 to satisfy the limit of validity for the laws of the wall.

Cyclic boundaries Cyclic boundaries consist of pairs of geometrically identical boundaries, where it is assumed that the gradients perpendicular to the segment plane are determined from the flow field. This method can be exploited to reduce the size of the computational domain. This boundary type can be applied for swirling flow being symmetric to the z-axis and identical in all axial planes containing the symmetry axis. The segment angle φ is used to calculate the cross-velocity components v and w for the two segment surfaces. The transformation of the cross velocity components for both cyclic boundaries (1 and 2) is defined as

$$\begin{pmatrix} v_1 \\ w_1 \end{pmatrix} = \begin{pmatrix} \cos\varphi & \sin\varphi \\ -\sin\varphi & \cos\varphi \end{pmatrix} \cdot \begin{pmatrix} v_2 \\ w_2 \end{pmatrix} \quad \text{and}$$

$$\begin{pmatrix} v_2 \\ w_2 \end{pmatrix} = \begin{pmatrix} \cos\varphi & -\sin\varphi \\ \sin\varphi & \cos\varphi \end{pmatrix} \cdot \begin{pmatrix} v_1 \\ w_1 \end{pmatrix} \quad (5.34)$$

The scalar quantities, such as pressure and density and the axial velocity u, are identical for corresponding points at the two segment surfaces. The mapping of the different variables from one surface to the other can be done in an arbitrary or non-arbitrary way, depending on the type of grid (unstructured or structured).

Slip surface boundary The slip surface boundary offers no resistance to tangential motion and therefore represents a surface without friction.

5.7 Grid Sensitivity Analysis

The number and density of control volumes in the computational domain influences the accuracy of the spatial resolution of the occurring flow effects. Additionally, the number of cells also directly influences the computational effort and the necessary memory capacity. A sensitivity study with different computational grids is performed to evaluate the dependency of the results from the analysis for the interesting variables. The cell size

5.7 Grid Sensitivity Analysis

within the computational domain is refined in five steps. The refinement, and therefore the number of cells, ranges from a coarse grid with 293077 cells up to 1103708 cells for the finest grid. To satisfy the validity for the laws of the wall for each model, the cell layer next to the wall is kept inside the limit for the y^+-value, while the progressing cell layers into the flow domain are adapted to the refinement. The dominant variable for the evaluation and optimization of the backflow limiter for normal operation condition is the local pressure drop for each section and therefore the overall pressure drop inside the component. Flow effects, which strongly influence the pressure drop like recirculation zones and separation, have to be resolved correctly by the chosen refinement. For the evaluation of the local and overall pressure drop several cutting planes along the z-axis and in radial direction are necessary. It is not possible to quantify the values directly from the local planes due to the unstructured character of the mesh. In order to weight the local pressure values stored inside the node at the cell volume center of each cell which are cut by the evaluation plane, a Gaussian filter function w of the following form is applied to average the volume of cells:

$$w = e^{-g(x-x_i)^2} \tag{5.35}$$

where g is a parameter that controls the width of the averaging volume of cells that is cut out of the model at the position x_i. Depending on the size of the fluid cells in the vicinity of the cutting plane and therefore the distance of the cell center node relative to the cutting plane, the factor g has to be adapted to give a representative mean value for the variable. The x coordinate represents a local coordinate perpendicular to the cutting plane, which quantifies the distance between the nodal cell center of an individual cell to the center of the global coordinate system along the axis perpendicular to the cutting plane. x_i defines the distance of the cutting plane to the center of the global coordinate system along the axis perpendicular to the cutting plane. The weight factor reaches unity when the center node is directly located on the cutting plane and decreases to zero for big distances of the center nodes from the cutting plane. The general expression for the pressure p at a position x_i can be written in cylindrical coordinates as:

$$\Delta p = \sum_{i=2}^{n} (p(z_{i-1}) - p(z_i)) \tag{5.36}$$

The overall pressure loss Δp is then defined as:

$$\Delta p = \sum_{i=2}^{n} p(z_{i-1}) - p(z_i) \tag{5.37}$$

where n represents the number of cutting-planes along the flow path. The same method is also used to determine the swirl S for the optimization. To estimate the dependency of the overall pressure drop on the cell size of the computational grid, the mean cell length of the individual refinement models is determined by dividing the volume of the whole fluid domain by the amount of cells inside this volume and applying the cubic root to the calculated mean cell volume. The first order interpolation scheme UD and the third order interpolation scheme QUICK for the spatial discretization have been tested for the five different refinement models. In Figure 5.2, the overall pressure loss Δp for the five different grids and the two interpolation schemes is plotted over the average edge length of the mean cell volume for each grid. The results for the applied UD-scheme (blue

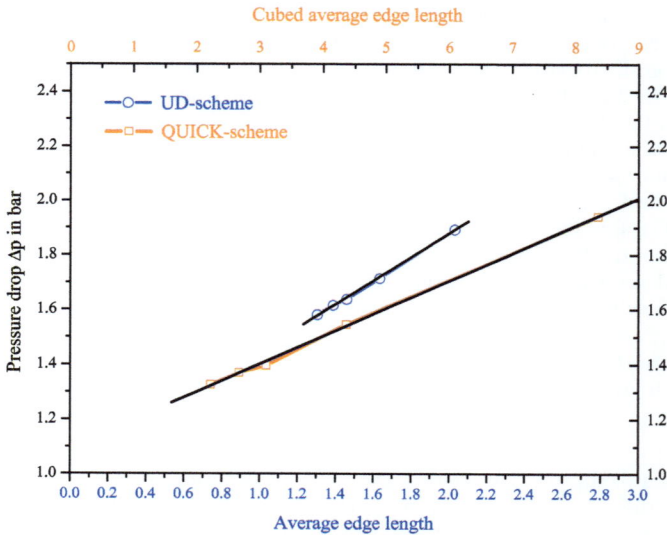

Figure 5.2: Grid convergence study using five different mesh topologies for the UD- and QUICK-scheme showing the evolution of the overall pressure drop in the backflow limiter for normal operation condition.

triangles) decrease in a linear manner as expected for a first order scheme, while the QUICK-approach (red squares) yields a cubic decrease due to its third order accuracy. The numerical error for the different grids and interpolation schemes can be approximated using the Richardson-extrapolation as defined by Roache [71]. The values of the pressure drop for the different edge lengths are extrapolated to the y-axis which represent the edge length of an infinite fine mesh. For the UD-scheme a pressure drop of 1.01 bar can be extrapolated, for the QUICK-scheme a slightly higher pressure drop of 1.09 bar is determined. The QUICK-scheme is chosen for the calculations, it yields the lowest error with its third order accuracy. With the available computational power, the maximum possible resolution with over 1.1 million cells is chosen, which has an estimated error of 17.4 % compared to the extrapolated value from the Richardson extrapolation.

5.8 Evaluation of Turbulence Models

Three different eddy viscosity turbulence models have been tested to determine their influence on the important variables in the flow: the linear standard k-ε, the ω-SST and the non-linear quadratic k-ε. Several investigations can be found in the literature for incompressible fluid applications involving separation zones and swirl in ducts and water turbines (ERCOFTAC [15], MARNET-CFD [98], QNET-CFD [14]). The ERCOFTAC Best Practice Guidelines recommend the use of the ω-SST model for a better prediction of flow separation from surfaces under the action of adverse pressure gradients. For swirling

5.8 Evaluation of Turbulence Models

flows, non-linear eddy viscosity models are recommended, requiring more computational effort as a back draw.

Investigations have also been performed by Casey [14] inside the QNET-CFD framework. Among several other flow regimes, a geometry with varying cross sections has been investigated by Cervantes and Engström [16]. They found, that the ω-SST model predicts the flow distribution and pressure variations quite accurately compared to the experiments. This model seems promising for the flow analysis of the backflow limiter, since it exhibits a number of cross section variations in the axial part to influence the pressure distribution for a minimum pressure loss.

For the backflow limiter, the steady-state evaluation is performed adopting the SIMPLE algorithm and the AMG method to solve the discrete equations, as spatial discretization method the QUICK scheme with a convergence criteria of 10^{-4} is applied. The individual pressure loss for each section I to VIII of the backflow limiter (see Figure 3.1, page 27) is determined for the investigated turbulence models and shown in Figure 5.3. It can be stated that the ω-SST and the quadratic k-ε predict an almost

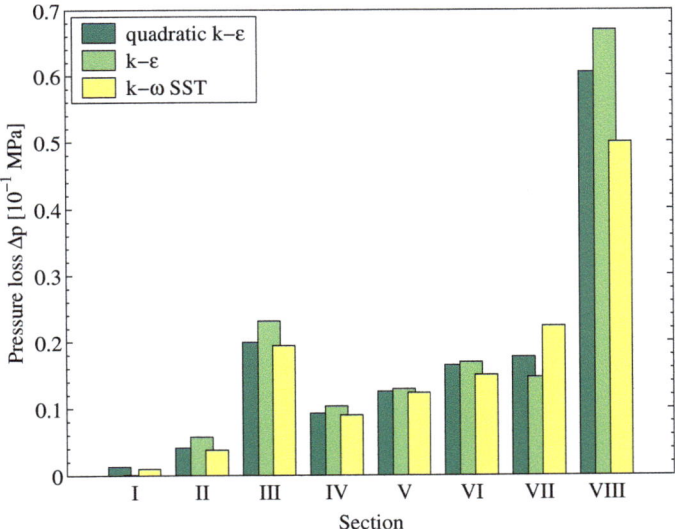

Figure 5.3: Comparison of the individual pressure losses Δp for sections I to VIII of the backflow limiter for the k-ε, ω-SST and quadratic k-ε model.

identical pressure loss for the sections I to V, which represent inlet section, the inlet nozzle, the inlet swirler vane, the axial nozzle and the axial diffuser. In those regions the flow does not display separation zones for all three models, the higher pressure loss for the k-ε model can be explained with the overprediction of the turbulent kinetic energy compared to the other two. For the axial diffuser section the pressure losses of the k-ε, the quadratic formulation and the ω-SST model have the same pressure loss, which can be explained by the dominant effect of pressure recovery, superposing the other effects.

For the radial diffuser, the boundary layer modeling is the most influential parameter due to the large and curved wall area, which can be predicted more accurately by the ω-SST model. The lower pressure loss for the SST-model shows that the optimized design omits separation in this zone.

For the swirl chamber, the same effect is responsible for the in this case higher pressure loss due to the strongly expanding cross section in radial direction for the ω-SST model compared to the other two. The pressure loss for the exit swirler vanes is strongly influenced by the velocity profile, especially of the outward velocity vector in relation to the exit blade angle. Here, the three models show deviating pressure losses, with the ω-SST model exhibiting the lowest pressure loss, followed by the quadratic and the linear k-ε model. The non-linear k-ε model predicts the swirl distribution the best, being an important, but not dominant factor for the pressure loss optimization in normal operation condition.

For the evaluation of the overall pressure loss, the ω-SST model is still closer to the quadratic model then the linear k-ε model. As a summary, it can be stated that the ω-SST model is suitable for the regions, where the cross section variation and the wall effect is dominant for the pressure loss, while the non-linear k-ε model is superior for regions where the swirl dominates the pressure distribution in the flow.

Part III

Concepts and Analyses

6 Reactor Design Overview

A short overview of all three reactor concepts is given in the following chapter. Table 6.1 summarizes the characteristics for the investigated core concepts. Main data for the one pass core concept as presented by Vogt et al. [93] include a mass flow rate of 2810 kgs^{-1} and heat-up from a feedwater temperature of 280 °C to 380 °C. For this moderate heat-up of around 100 °C in the core, peak cladding temperatures of 416 °C can be expected, allowing for stainless steel as cladding material. An overall efficiency of 37.5 % can be achieved, which is about 2 % higher compared to conventional PWR designs.

	One pass	Two pass	Three pass
Core characteristics			
according to	Vogt [93]	Kamei [45]	Schulenberg [78]
Feedwater temperature [°C]	280	280	280
Steam exit temperature [°C]	380	500	500
Flow rate [kgs^{-1}]	2810	1418	1160
Specific power [MWm^{-3}]	100	59.9	57.9
Peak cladding temperature [°C]	416	650 [1]	620 - 630
P_{th}/P_{el} [MW]	2000/750	2744/1200	2400/1000
Net efficiency	37.5	43.8 [2]	44
Design			
Number of FA clusters	88	88	156
Inner diameter vessel [m]	3.38	3.38	4.47
Maximum shell thickness	0.51	0.51	0.56
Height [m]	13.3	13.3	14.3
System pressure [MPa]	25	25	25

Table 6.1: Characteristics of the three different reactor designs.

The two pass core concept has not yet been investigated for the HPLWR; data are only available for the Super LWR two pass core concept introduced by Kamei et al. [45]. In their investigation, a fuel assembly design with 300 fuel rods, 36 square water rods, and 24 surrounding rectangular water rods has been chosen. The core comprises 121 of these clusters yielding a thermal power of 2744 MW$_{th}$ with expected steam outlet temperatures around 500 °C. The local peak cladding temperature has been determined to 650 °C not yet including uncertainties and allowances. The expected higher temperatures in the vicinity of 700 °C exceed the limits of available cladding materials.

Schulenberg et al. [78] proposed a three pass core concept for the HPLWR which includes steam exit temperatures of 500 °C resulting in a net efficiency of 44 %. For a

---
[1]Published results do not include power peaking due to allowances and uncertainties.
[2]Overall thermal efficiency, net efficiency has not been reported.

flow rate of 1160 kgs^{-1} and a thermal power of 2400 MW$_{th}$ the local peaking temperature has been determined to be around 620 to 630 °C in their analysis. Stainless steel claddings can be applied in this case, as power peaking for allowances and uncertainties is already included for the investigation.

The internals and design details of the reactor pressure vessel are using state of the art technologies of pressurized water reactors, as far as possible, such as core barrel with its core support plate and alignments, the steel reflector, the CRGA and its guide tubes, the vessel closure head with its sealing and the vessel bolt design. Other components have to be designed differently to account for the higher steam outlet temperature. The RPV differs from a conventional PWR by a larger wall thickness to stand the higher system pressure of around 25 MPa. Depending on the core design with one, two or three passes the inner diameter of the vessel varies. Components such as the upper steam plenum, the lower mixing chamber and the fuel assembly clusters have to be modified for each core design to realize the proposed flow path. The necessary modifications for the fuel assembly cluster of the two pass core have been published on the 28th Annual Conference of the Canadian Nuclear Society in 2007 [75]. The design of the fuel assembly for a three pass core and the corresponding upper and lower mixing plenums has been presented at the Global conference in 2007 by Fischer et al. [28].

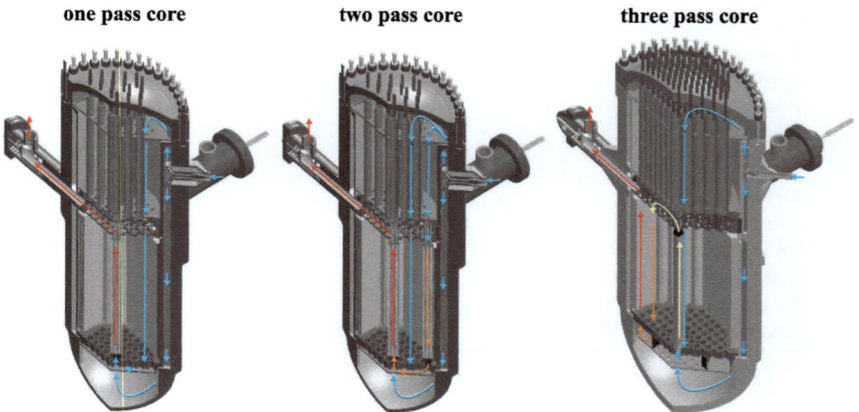

Figure 6.1: Illustration of all three reactor designs with indicated flow path. Cold feedwater enters the inlet (indicated with blue colored arrows), and is heated in one or several steps (indicated with yellow, orange and red arrows) depending on the core design.

A detailed description of all components of each RPV assembly can be found in Appendix B; Figure 6.1 illustrates the RPV design for the one, two, and three pass core respectively. The first two variants feature 88 fuel assembly clusters resulting in an inner vessel diameter of 3.38 m. The design of the RPV and the internals for both, the one and two pass core has been presented by Fischer et al. [30] and [29]. The executed verification of the design of the RPV with thermal sleeve using a finite elements approach can be found in Fischer et al. [24].

To realize the three pass core concept as proposed by Schulenberg et al. [78], a total of 156 fuel assembly clusters is necessary, yielding an increased inner vessel diameter of 4.47 m. The maximum shell thickness of the two smaller vessels reaches 0.51 m in the vicinity of the upper flange, using a conventional flange design. The vessel with the larger diameter has been further optimized with a maximum thickness of 0.56 m, to stay inside the forging limit for pressure vessels.

All designs feature four inlets and four outlets with an increased diameter of the outlet to compensate for the increased volume of the coolant. The height of the three pass core vessel is increased from 13.3 m to 14.3 m compared to the other two designs, which can be accounted to the more complex design of the fuel assembly cluster and the two plenums. The design characteristics of all three concepts are summarized in Table 6.1. The flow path of the coolant is indicated in Figure 6.1 for each design, where the cold feedwater enters the inlet (indicated with blue colored arrows), and is heated in one or several steps (indicated with yellow, orange and red arrows) depending on the core design. Table 6.1 also features the core characteristics of each design. The parameters of the one and three pass core have been directly applied for the design of the pressure vessel and internals. The two pass core values are extracted from the Super LWR design presented by Yamaji et al. [99] and are not directly applicable for the design of the HPLWR two pass core itself and the resulting design of the RPV and internals. They shall just give an indication of the possible parameters using a two pass core arrangement for a supercritical water-cooled reactor.

7 Design of Core Components

7.1 One Pass Core

The one pass core of the HPLWR features the assembly design presented by Hofmeister et al.[41] with clusters of 9 assemblies with 40 fuel pins each. A short review of the design is presented here since the fuel assembly cluster design for the two and three pass core is derived from this design. The dimensions of the assembly box, the moderator box and the fuel rod can be found in Table 7.1. The width of the outer assembly box is only 67.2 mm, which is about one third of the typical fuel assembly size of a PWR. Nine of these small assemblies are clustered in a 3×3 arrangement to use the conventional control rod drive of a current PWR.

Component	Dimension [mm]
Fuel rod	
Cladding outer diameter	8
Cladding thickness	0.5
Active height	4200
Upper/lower fission gas plenum height	255
fuel rod total height	4710
Assembly box	
Inner size	65.2
Wall thickness	1
Inner corner radius	5
Axial length	4841
Water gap between assemblies	10
Moderator box	
Outer size	26.8
Wall thickness	0.4
Outer corner radius	4.2

Table 7.1: Fuel assembly cluster dimensions for the one pass core design as described in Hofmeister et al. [41].

The design allows a leak-tight counter current flow of moderator water and coolant, which is necessary to provide a uniform axial power profile in the core. Moderator water is flowing downwards through boxes inside the assemblies and through the gaps between the assembly boxes (see Figure 7.1), whereas the coolant rises upwards.

The head piece consists of four components: a head piece plate, a transition nozzle, a window element and a bushing. The nine assembly boxes are welded into the square plate which in turn is welded with the transition nozzle and the window element. The transition nozzle reduces the square cross section of the cluster to a smaller, round cross section above to apply two metal C-rings ensuring leak tightness with the steam plenum. The window element releases the hot steam horizontally through four windows into a steam plenum mounted over all head pieces. The moderator boxes and tubes are welded into the top of the head piece to supply moderator water for the assemblies. Control rod spiders featuring five cruciform fingers are inserted into the square moderator boxes from the core top like in a PWR. The bushing, which is screwed onto the window element, enables handling of the cluster for maintenance and replacement.

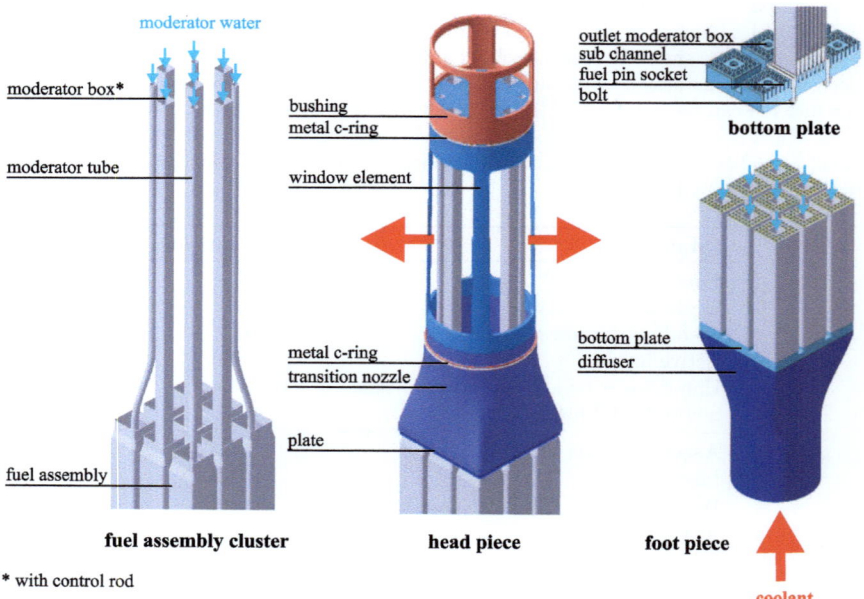

Figure 7.1: Fuel assembly cluster head piece and foot piece design by Hofmeister [41] with indicated flow path of the coolant.

The foot piece, which is shown in Figure 7.1 on the right side, comprises a bottom plate and a diffuser. The 40 fuel rods of each assembly are inserted into the sockets of the bottom plate. Only the central assembly box is connected with the foot piece using four bolts. The other eight assembly boxes are hanging freely from the head piece to minimize thermal tensions within the cluster. The moderator boxes and tubes are equipped with outlet nozzles fitting into the outlets of the bottom plate. The foot piece has been optimized such that it mixes homogeneously the moderator water released from moderator boxes and tubes, from the assembly gaps and additional feed water, supplied through the downcomer directly to the foot piece inlet. An inlet orifice underneath the foot piece adjusts the coolant mass flow to the cluster power.

Disadvantages of this design include a complicated square shaped sealing between bottom plate and assembly boxes for the foot piece. The design of the bottom plate itself is very complex and complicated to manufacture. Additionally, high pressure losses are expected for the coolant due to the small subchannel inlets.

7.2 Modifications for a Two Pass Core

For the two pass core concept as proposed by Yamaji et al.[100], Schulenberg et al. [74], [75], the coolant must be heated up in steps with intermediate mixing to avoid overheated fuel rods in hot channels. A possible core design with a two-step heat up uses coolant running downwards in peripheral fuel assemblies being preheated already to around 380 °C in the lower plenum. From there, it rises in the inner fuel assemblies of the core where it is heated up to core exit temperature of around 430 °C. These exit temperatures allow to match the creep and corrosion limits of stainless steel. The inner fuel assemblies serve as a superheater of the core described above. Again, moderator water flows downwards in moderator boxes to compensate for the high density change in the core for a uniform power profile. Therefore, two different flow regimes are possible for the fuel assembly clusters. For the co-current flow in the peripheral clusters, coolant flows downwards with the moderator water in the moderator boxes. In the case of the counter-current flow for the inner assemblies, coolant rises in the core while the moderator flows downwards.

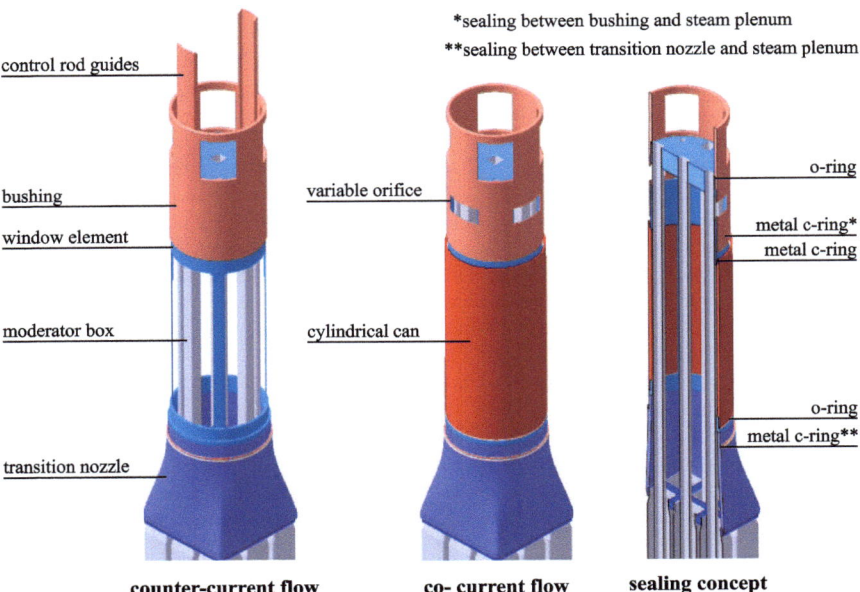

Figure 7.2: Fuel assembly cluster for counter-current (left side) and co-current coolant flow (middle) with modified head pieces for the two pass core and applied sealings (right side) as described in Schulenberg et al. [74], [75].

Using the assembly design of Hofmeister, there are only a few modifications needed to enable such a flow path. While the foot piece remains unchanged, the head piece must be extended about 100 mm above the upper plate of the steam plenum, as shown in Figure 7.2. Two different bushings, screwed onto the head piece and sealed with an o-ring, will be needed. In case the coolant should flow downwards (co-current flow), a bushing with inlet openings for feedwater in combination with a long cylindrical can is screwed onto the head piece. The cylindrical can is used to close the exits of the window element for the steam and is sealed using an o-ring at the bottom and a C-ring at the top (Figure 7.2 to the right side). The size of the inlet openings in the bushing adjusts the individual mass flow rate needed for the cluster. In the case, when coolant flows upwards to be released to the steam plenum (counter-current flow), the cylindrical can is removed and a closed bushing is screwed onto the head piece instead. With such a cluster assembly design, any position in the core can be selected for downward flow, not only the peripheral ones. Each assembly can be used either for upward or for downward flow, and the flow direction of a cluster can simply be changed during a revision by exchanging the can and the bushing.

Special attention has to be payed to the mixing of coolant between the subchannels for upwards and downwards flowing coolant. Adequate measurements have to be taken to allow a coolant flow in both directions with a comparable high mixing.

7.3 Three Pass Core

Higher core exit temperatures, and thus a higher specific turbine power and a higher net efficiency, can be achieved by following the concept of supercritical fossil fired boilers including a second superheater. The resulting three pass core concept has been sketched by Schulenberg et al. [78]. The part of the core in which water is changing its density from liquid-like to steam-like properties is called evaporator, even though this transition does not show any boiling phenomena. The evaporator assemblies allow the highest power of fuel pins due to their larger temperature margin to cladding material limits than superheater assemblies. They are placed in the center of the core, accordingly, where the neutron flux is highest. Underneath its inlet at the core bottom, all moderator mass flows from moderator boxes and from gaps between assemblies are mixed with feedwater from the downcomer to an inlet temperature of around 310 °C. The evaporator heats up the coolant to 390 °C.

An inner mixing chamber above the core eliminates hot streaks. Two sets of superheater assemblies are arranged concentrically around the evaporator, the first one having a downward flow of coolant starting from the evaporator exit, and the second one having an upward flow again. This way, the second superheater, which has the highest coolant temperatures, is at the core periphery where the neutron flux and the pin power are lowest, such that peak cladding temperatures in the superheaters are similar to those in the evaporator. The first superheater with downward flow heats the coolant up to 433 °C. After a second mixing in an outer mixing chamber below the core, the coolant will finally be heated up to 500 °C with upward flow in the second superheater in peripheral assemblies.

The mechanical core design presented here shall use the same assemblies for all heat-up steps to allow the highest flexibility for any assembly repositioning to optimize the radial

7.3 Three Pass Core

power profile and burn up. The assembly design is based on the Hofmeister design. The proposed flow path, however, requires some modifications to this design.

First, all moderator water may only be mixed with the coolant upstream of the evaporator inlet and not in the foot pieces of superheater assemblies. Any addition of cold moderator water to the superheated steam would cool it down, creating cold and hot streaks which need to be avoided to minimize peak cladding temperatures. Therefore, the assembly design must be tight against leakage of moderator water.

Secondly, spacers between the fuel rods shall mix the coolant for any flow direction, upward or downward. Wire wraps, as used in liquid metal cooled reactors (see Hoffmann [38], 1973), are among the most promising options for this application.

Third, a multi pass flow through the core will cause a significantly higher pressure drop of the coolant which can partly be mitigated with a slightly higher pitch of the fuel rods and with minimized pressure losses in head and foot pieces of the clusters. As recirculation pumps are not foreseen for this reactor concept, a higher pressure drop will only affect the power of the feedwater pumps, such that the pressure drop is far less critical for cycle efficiency than in a BWR. A remaining disadvantage, however, will be a higher pressure difference across assembly and moderator boxes, requiring thicker box walls.

Finally, the coolant should be mixed homogeneously before each heat-up step to minimize the peak cladding temperature in each assembly. This requires special attention to the design of a coolant plenum above and below the core. They should be designed as efficient mixing chambers.

Component	Dimension [mm]
Fuel rod	
Cladding outer diameter	8
Cladding thickness	0.5
Wire diameter	1.34
Wire axial pitch	200
Active height	4200
Upper/lower fission gas plenum height	255
fuel rod total height	4710
Assembly box	
Inner size	67.5
Wall thickness	2.5
Inner corner radius	5.44
Axial length	4851
Water gap between assemblies	10
Moderator box	
Outer size	26.9
Wall thickness	0.8
Outer corner radius	4

Table 7.2: Fuel assembly cluster dimensions for the three pass core design.

The mechanical core design data used here are listed in Table 7.2. A total number of

40 fuel rods per assembly with 8 mm outer cladding diameter at a pitch of 9.44 mm, are housed within a stainless steel box of 2.5 mm wall thickness and 72.5 mm outer size. A single wire of 1.34 mm diameter is wrapped around each fuel rod with an axial pitch of 200 mm, leaving a tolerance of 0.1 mm between the wire and the fuel rods or the box walls, respectively. The inner moderator box has an outer size of 26.9 mm and a wall thickness of 0.8 mm. It is made from stainless steel as well. An active core height of 4.2 m is assumed. Including the fission gas plena, the fuel rods reach a length of more than 4.7 m. The assembly cluster, built with 9 assemblies and with a gap of 10 mm between the boxes, is shown in Figure 7.3.

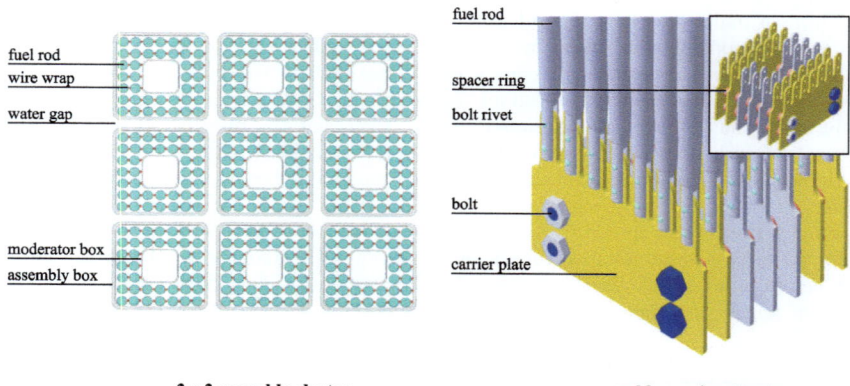

3 x 3 assembly cluster   **assembly carrier structure**

Figure 7.3: Cluster of 9 assemblies with a central moderator box and 40 wire wrapped fuel rods each (left side), and corresponding carrier structure of the fuel rod bundle (right side) for the three pass core design.

Up to seven fuel rods are bolted with a carrier plate at their lower end, as shown in Figure 7.3 to the right. As the central moderator box cuts out 3 fuel rods of this array, three inner carrier plates (shown in gray in Figure 7.3) have only 4 fuel rods each. All carrier plates of an assembly are connected with 4 bolts and spacer rings, which can be disassembled if a fuel rod should be exchanged. This pre-assembled fuel rod bundle is inserted into the assembly box from the foot piece side. The upper end of the fuel rods is freely expanding and horizontally positioned only by the wire wraps and box walls. The pressure loss of these support structures are expected to be significantly smaller than the pin socket structure proposed by Hofmeister (see Figure 7.1).

The foot piece is designed with an upper plate, a lower plate, and a diffuser (which becomes a nozzle in case of the first superheater), as shown in Figure 7.4. All but the central assembly box of the cluster are welded with the upper plate. The central assembly box is bolted with 4 assembly box screws of the size M10 with the upper plate, instead. This central box is carrying the cluster weight when the cluster is lifted up. The other 8 assembly boxes are sliding with their upper ends in the head piece to avoid thermal deformations of the cluster in case of different assembly temperatures (see Figure 7.5, right side). All central moderator boxes inside the assemblies are welded to the top of the head piece. Their lower ends are extended with cylindrical tubes which are inserted

7.3 Three Pass Core

into the lower plate of the foot piece, where the moderator water is released to be mixed with the gap water surrounding the assemblies (indicated flow path with blue arrows in Figure 7.4).

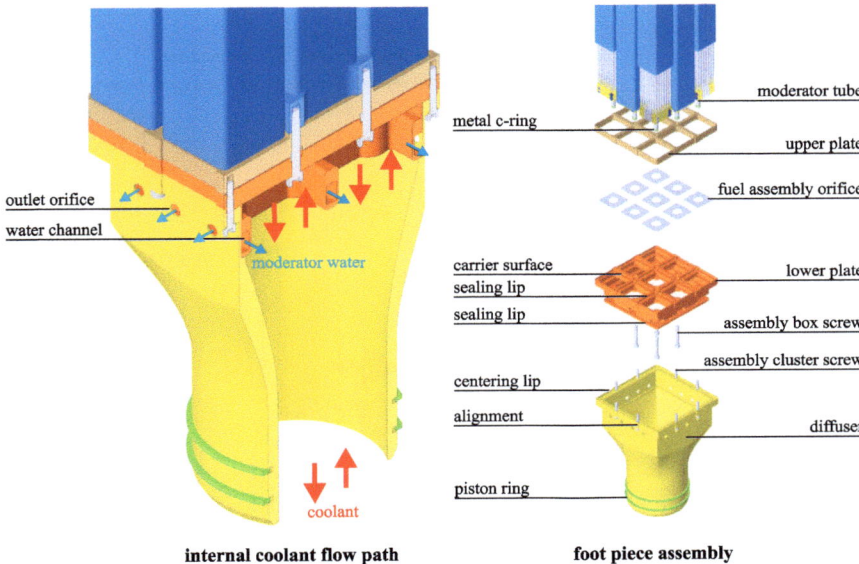

Figure 7.4: Foot piece of the three pass core fuel assembly cluster (right side) and indicated flow path (left side).

These holes can be used as outlet orifices to adjust the mass flow rate through the moderator boxes. Openings for the vertical steam flow, which are surrounded by the lower plate, are designed as large as possible to minimize pressure losses. The lower plate of the foot piece and the diffuser are welded together to avoid leakage of cold moderator water into the superheated steam. The diffuser features circumferentially arranged alignments to position the cluster in the core, and to prevent rotational movement. After insertion of the fuel rod bundles into the assembly boxes, the lower plate and the upper plate of the foot piece are bolted together with 8 assembly cluster screws of size M8. This way the lower plate keeps the fuel bundles in place. The upper plate and the lower plate have sealing lips at their contact surfaces which are pressed together by the assembly cluster screws. The total arrangement is completely separating moderator and steam mass flows without any significant leakage.

If necessary, a fuel rod can be replaced during revisions e.g. for inspection with the following disassembly steps. The cluster is turned upside down, the 12 bolts of the foot piece are opened, and the fuel rod bundle is pulled out. After opening the 4 bolts of the carrier plates, each row of fuel rods can further be disassembled by drilling out the bolt rivets of the dedicated rod (see Figure 7.3, right side). The cluster is reassembled in reverse order.

The head piece of the assembly cluster consists of a bottom plate with nine round openings for the cylindrical extensions of the assembly boxes, a transition nozzle between the square cluster geometry and the cylindrical openings of the steam plenum, a cylindrical window element with large openings to release the steam, and a bushing screwed on top of the head piece to pull out the cluster. The head piece is shown in Figure 7.5.

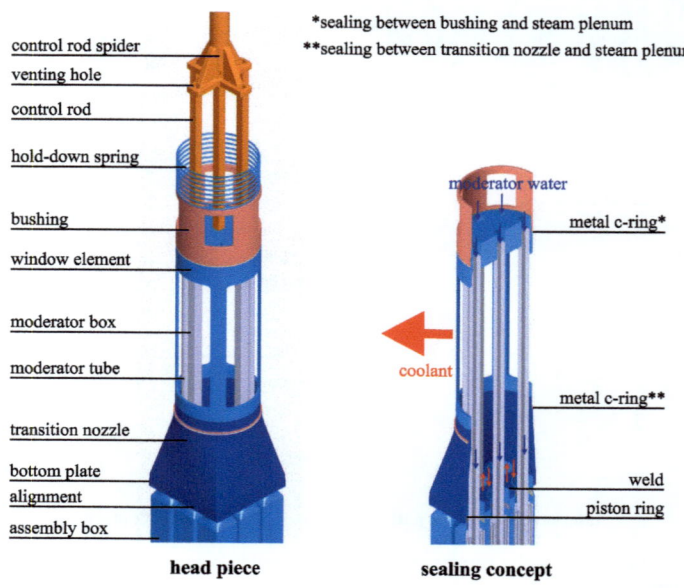

Figure 7.5: Head piece of the three pass core fuel assembly cluster with inserted control rod spider and hold-down spring (left side) and sealing concept (right side).

All but the inner assembly box are sliding in the bottom plate of the head piece. Two piston rings, shown in Figure 7.5, avoid leakage there. The central assembly box is welded with the bottom plate. The bottom plate, the transition nozzle and the window element are welded together as well. Two metal C-rings are applied to the window element to minimize leakage of moderator water into the steam plenum, which has already been described for the one pass core head piece. The extensions of the moderator boxes, of which four are round to be bended to fit into the window element, are welded into the top plate of the window element. The other five moderator boxes have straight extensions to allow insertion of control rods from the top.

All assembly clusters are resting with their diffusers on a core support plate similar to the one pass core design, as can be seen in Figure 7.6. Two piston rings (see Figure 7.4) around each diffuser avoid leakage of colder moderator water, coming from the gaps between the assemblies, into the superheated steam of the outer mixing chamber. This moderator water is released instead to the inner mixing chamber at the evaporator inlet through openings in the core support plate. The remaining feedwater, supplied through the downcomer surrounding the core, is injected into the same inner mixing chamber

7.3 Three Pass Core

from below, providing an effective turbulent mixing there.

Component	Dimension [mm]
Steam plenum	
Outer diameter	3965
Maximum diameter inner mixing chamber	2956
Inner height	480
Wall thickness horizontal plates	60
Wall thickness peripheral shell	30
Diameter head piece opening	218
Outer diameter connection tube	82
Wall thickness connection tube	5
Lower mixing plenum	
Outer diameter	3533
Maximum diameter inner mixing chamber	1961
Inner height	550
Wall thickness horizontal plate	15
Wall thickness peripheral shell	15

Table 7.3: Steam plenum and lower mixing plenum dimensions for the three pass core design.

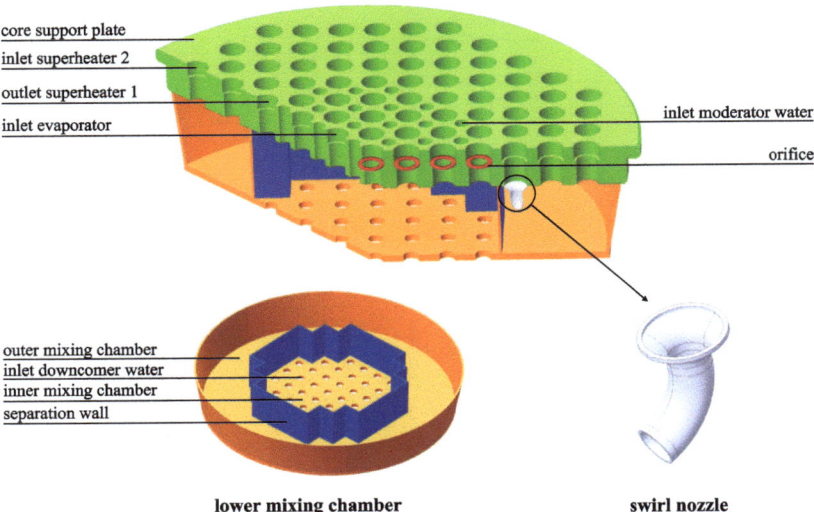

Figure 7.6: Lower mixing chamber with core support structure and swirl nozzle for the three pass core fuel assembly cluster.

Similarly, mixing of superheated steam in the outer mixing chamber can be enhanced by horizontal swirl nozzles at the outlets of the superheater 1, shown exemplary in Figure

7.6 on the lower right side, causing a turbulent swirl there. The effectiveness of jets for turbulent mixing has been demonstrated by Hofmeister et al. [39] in a similar context.

Dimensions of the lower mixing plenum can be found in Table 7.3. After installation of all assembly clusters of the core, the steam plenum, shown in Figure 7.7, is mounted over all window elements using protruding guide strips. It consists of two horizontal plates, each with 60 mm thickness, leaving a steam volume with 480 mm height between them (see Table 7.2 for details). Connection tubes between both plates guide some moderator water to the gaps between the assemblies and ensure sufficient stiffness of the design. Vertical walls, shown in Figure 7.7, are welded between some of the connection tubes. They separate an inner mixing chamber, where the coolant is mixed between the evaporator outlet and the superheater 1 inlet, from an outer mixing chamber which collects the superheated steam to exit through the nozzles and be supplied to the high pressure steam turbine. Four so-called hot tubes to direct the steam outside can be driven radially into the outer mixing chamber through the exit nozzles. Orifices at the inlets of the connection tubes adjust the moderator mass flow rate through the assembly gaps. These design details illustrate how a leak tight, multiple flow path of the coolant through

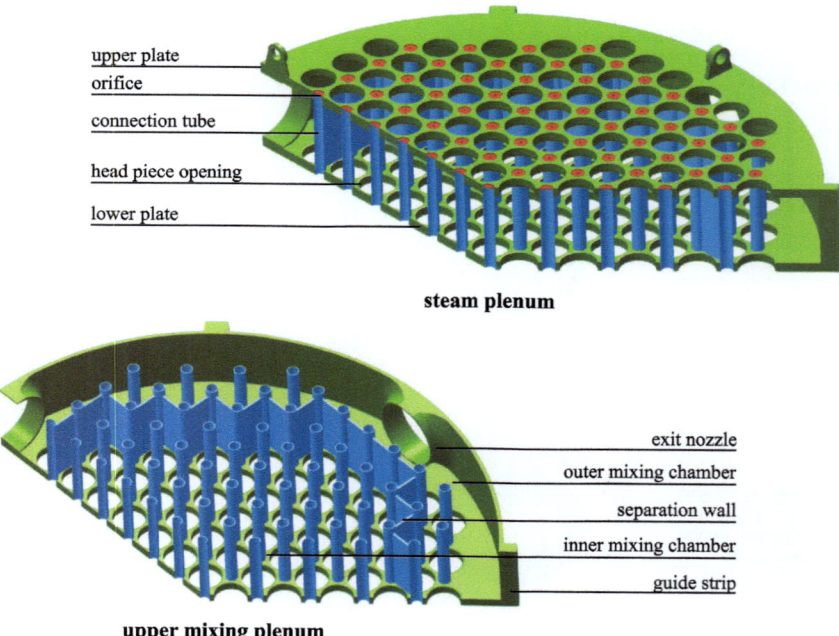

Figure 7.7: Steam plenum with two mixing chambers and integrated connection tubes for the three pass core design.

the core can be achieved with only a few modifications compared with a conventional, single flow path. For both, the two pass core with its co-current and counter-current flow arrangement and the three pass core with evaporator and both superheaters, the

7.3 Three Pass Core

total coolant flow path remains closed against leakage of colder moderator water even in case of larger thermal expansions of the components. Moreover, all mass flows through water tubes and through gaps between the assembly boxes are adjustable, leaving enough flexibility for later optimization. Assembly clusters can be exchanged freely between the different regions of the core for the three pass arrangement, and with slight modifications for the two pass core. This facilitates the optimization of the power profile and maximizes the burn up.

8 Internals Design

The core barrel is composed of a cylindrical part with flange and the support core plate with lower mixing plenum; this is shown in Figure 8.1. The principle design of the core support plate has already been presented by Hofmeister [41]. The purpose of the core barrel is the containment and fixation of the core with the fuel assembly clusters, standing directly on the perforated core support plate. This arrangement allows the clusters to maintain their orientation and position during operation. The upper part of the plate is formed like a shoulder and is welded to the bottom ledge of the core barrel. The radial outer surface of the lower part is used to radially align the core barrel inside the RPV. The

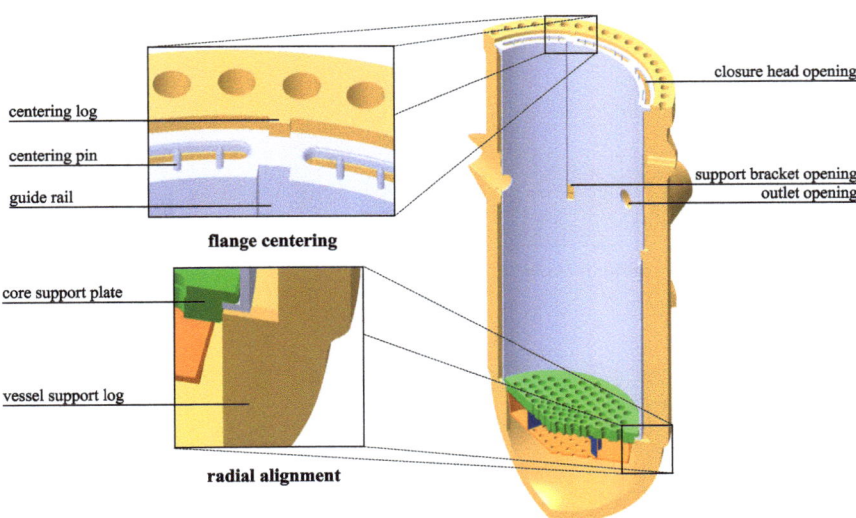

Figure 8.1: Core barrel with core support plate and lower mixing plenum for the three pass core design.

lower mixing plenum, which is welded to the bottom of the lower core plate, homogenizes the water flow from the downcomer before it enters through the plate into the lower part of the evaporator of the core. Here it mixes with the coolant from superheater 1 before it enters superheater 2. The purpose of the mixing plenum is discussed in more detail in chapter 7, page 66. Four slotted outlet holes are positioned at the height of the steam plenum to allow the connected hot tubes for the steam to pass through the cylindrical shell of the core barrel to the outlet flanges. Openings in the cylindrical shell allow to position the steam plenum on the protruding support brackets of the RPV. It is moved in and out of the cylinder using four stabilizing guide rails in the upper part. To enable coolant from the inlets to flow to the closure head dome, several gaps in the flange of the core barrel are provided. The arrangement and size of the gaps is selected to give a maximum coolant flow rate while having the highest possible stability for the flange at the same time. The flat top of the core barrel flange features pairs of centering pins, which are used together with the guide rails at the inner wall to center the CRGA inside the cylinder. It is desired to reduce neutron leakages from the core to protect the RPV

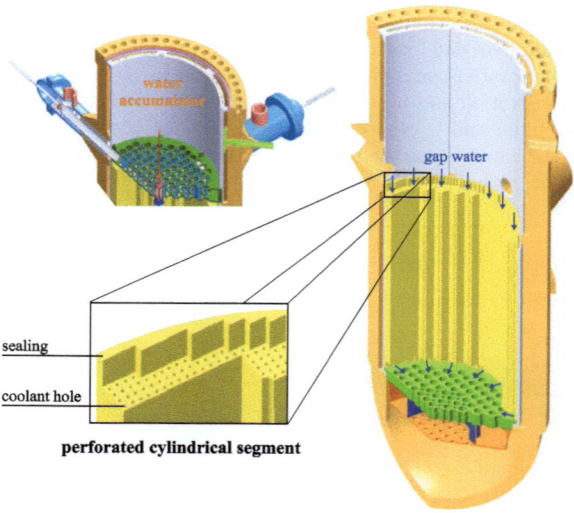

Figure 8.2: Steel reflector concept with sketched coolant flow path for heat removal for the three pass core design.

against fast neutron fluence-induced aging and to achieve a uniform power distribution.

Therefore, a steel reflector is introduced, filling the gap between the polygonal core and the cylindrical shell of the core barrel. It replaces the commonly used water reflector and thermal shield in PWR. This design is only preliminary and has not been optimized in a core analysis so far. Therefore, it gives one possible solution for the arrangement and the effective cooling of such a steel reflector. Additionally, the core design will strongly affect its dimensions. In the proposed design, the generated heat inside the structure is removed by gap coolant, flowing downwards from the upper plenum through the upper volume of the core barrel and through the connection tubes of the steam plenum inside the steel reflector. The heated coolant mixes at the bottom of the core with the remaining gap water (see Figure 8.2). Note that the illustrated steel reflector just shows a simplified design, to manufacture the structure with the displayed coolant holes, it is necessary to produce several smaller cylindrical rings. These segments are piled on top of each other and connected with bolts afterward. This design has already been implemented in the European Pressurized Water Reactor (EPR) by Areva NP [31].

The heated, supercritical steam is collected and mixed above all fuel assembly clusters in the steam plenum. This is a leak tight box, which sits on support brackets that are attached to the circumference of the vessel. Figure 8.3 shows a cross section of the RPV with outlets, the core barrel, steel reflector and the positioned steam plenum. For simplification, only one inserted fuel assembly cluster is displayed here. The collected supercritical steam leaves the steam plenum through one of the four hot pipes to be delivered to the high pressure turbine; the flow path is indicated in Figure 8.3 with red arrows. To replace spent fuel assemblies, the steam plenum is lifted out of the core barrel

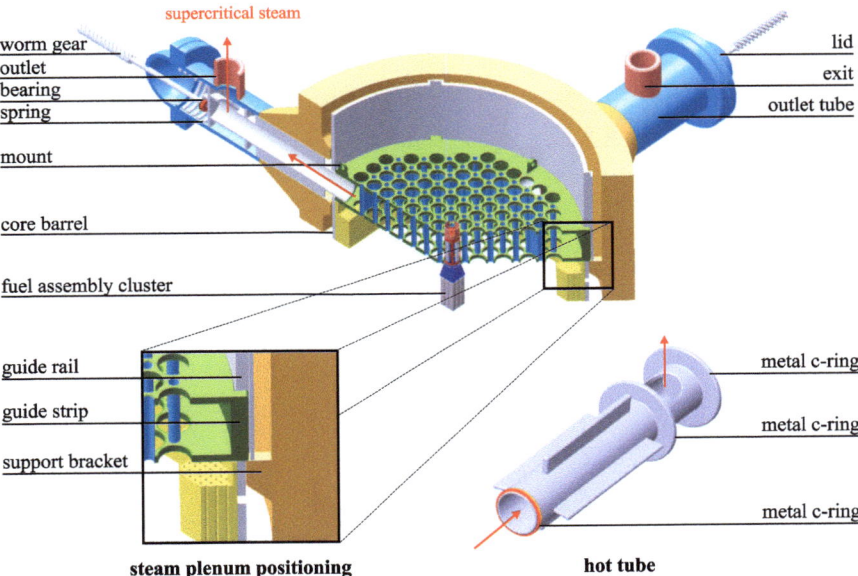

Figure 8.3: Steam plenum with extractable hot tubes and illustrated steam flow path for the three pass core design.

using four mounts welded to the top plate. For that purpose, the four hot tubes are moved radially outwards such that the nozzles of the hot tubes do not obstruct the lift path any more, but the steam plenum still rests on the protruding support brackets of the RPV. For a precise backward movement of the hot tubes, the pipe end is connected over a worm gear to an electric motor (not shown in Figure 8.3). Then, the steam plenum is moved out of the core barrel, and single fuel assembly clusters can be pulled out of the RPV. After replacing, the steam plenum is lowered and positioned with four guide rails at the inner wall of the core barrel. The counterparts at the structure for the alignment are four guide strips at the circumference of the steam plenum. The final position for the set down is defined by the support brackets of the vessel, where the bottom side of the steam plenum guide strip sits on. Then, the four nozzles of the hot tubes are moved inwards through the tubular openings of the casing.

To avoid leakage of cold water into the hot supercritical steam, the unavoidable gap between the hot tube and the steam plenum has to be sealed. The design features metal C-ring seals for that purpose (see Figure 8.3, detail on the right side), which are resistant to the hostile environment as described by Hofmeister et al. [40] and can be applied between moving parts. The c-shaped ring opens due to the pressure difference between the separated fluids and obstructs the gap between the two components, and therefore minimizes leakage. A safety ring at the end of the nozzle prevents the C-ring seal from sliding off when the tube is moved out of the steam plenum.

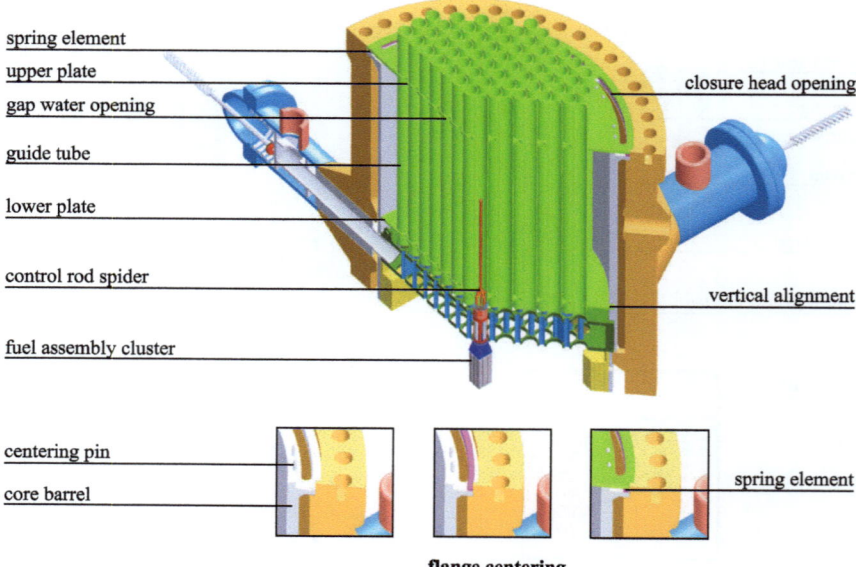

Figure 8.4: Control rod guide assembly (CRGA) assembled in the core barrel and the RPV for the three pass core design.

The design challenges in this case are the thermal expansions between the core barrel, the steam plenum and the RPV. If the steam plenum would be allowed to move with the expanding core barrel during operation, the fixed hot tube would jam inside the steam plenum and the C-ring seals would fail. Therefore, in this design, the steam plenum rests on the support brackets inside the lower vessel, while the core barrel is suspended at the closure head flange. With this concept thermal expansions between the internals and the RPV are decoupled and thermal stresses are avoided. The thermal expansions of the outlet pipes in the horizontal direction are compensated with a spring, implemented between the lid of the outlet tube and the inner hot tube. To seal the transition between the inner hot pipe and the exit of the outlet tube, two more C-ring seals are necessary between the two disc shaped barriers of the inner hot tube and the outlet tube. They are not shown in Figure 8.3 for simplification. Another design challenge results from the high temperature drop of over 220 °C between the inside and outside of the steam plenum, which causes very large thermal stresses and deformations in the structure. This issue is discussed in chapter 11.

The space above the steam plenum is dimensioned to house the guide tubes for the control rods. The tubes are connected to an upper and lower perforated support plate. The CRGA is suspended at the lower vessel top and centered inside the RPV using the centering pins attached to the upper surface of the core barrel flange; this is shown in Figure 8.4.

The upper support plate flange of the CRGA features the same gaps as the core barrel flange to allow coolant to flow to the closure head dome. The lower plate is centered inside the core barrel using the same guide strips as for the steam plenum. Both plates feature penetrations to allow gap water to fill the space between the guide tubes in the core barrel and then flow downwards through the connection tubes of the steam plenum to the core. This additional space for feed water above the core provides an accumulator in the case of loss of coolant accident (Hofmeister et al. [40]) and is illustrated in Figure 8.2 on the left side.

To ensure that the control rod maintains its position and orientation inside the guide tube, the housing inner wall is equipped with vertical alignment strips covering the whole length of the tube (Figure 8.5). Each guide tube is centered individually at the top of the steam plenum using the corresponding head piece bushing of the fuel assembly cluster. The bushing is designed in a way that the centering tube outer wall fits inside. A gap between the bottom of the housing and the upper side of the head piece (see detail in the lower right corner of Figure 8.5) allows for compensation of thermal expansions of the guide tube in horizontal direction and permits gap water to flow inside the several water boxes.

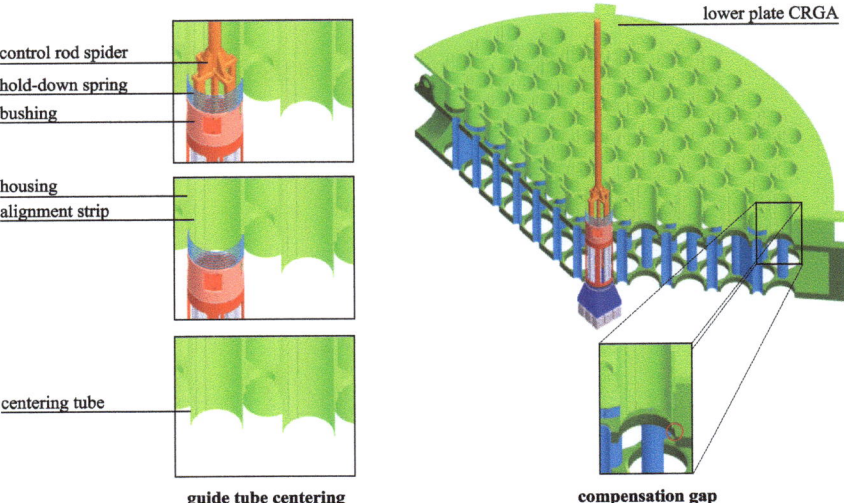

Figure 8.5: Guide tube centering in the head piece bushing for the three pass core design.

The different fuel assembly clusters tend to rise during operation due to buoyancy effects in the core. This causes damages either at the housing of the steam plenum, the guide tube or the transition nozzle of the FA-cluster itself. To avoid this problem, a spring is implemented, sitting on a ledge outside the guide tube housing and the upper edge of the bushing. It is not possible to mount the spring in the corner of the fuel assembly cluster, as in actual PWR, due to the limited available space for the very compact design of the core.

The number of guide tubes and therefore the number of control rods displayed in Figures 8.4 and 8.5 gives an example of one possible arrangement. The exact location and number of control rods has to be defined by core analysis.

A design for the HPLWR internals including the core barrel with core support plate and lower plenum, the steel reflector, the steam plenum with adjustable hot pipes and the CRGA is presented. This design combines several benefits, which are listed below:

- Steel reflector gives flattened power profile and reduces neutron leakages from the core compared to design with reflector water and thermal shield.

- Reduced thermal stresses and deformations between steam plenum and outlet tubes due to the presented component decoupling.

- Avoidance of feed water leaking into the steam plenum.

- Use of replaceable C-ring seals at the hot tubes and the steam plenum.

- Easy access to the core and the FA-clusters and therefore low complexity of loading/unloading procedure.

- In-vessel accumulator of coolant above the core giving additional safety in a loss of coolant accident.

- Design of the internals can be used for all three core designs with slight modifications only.

9 Reactor Pressure Vessel Design

The reactor pressure vessel (RPV) is designed to contain the three pass core and the internals. Therefore, the minimum height of the vessel is defined by the height of the fuel assemblies and the height of the extended control rods on top. In the radial direction, the diameter of the core and the thickness of the steel reflector and core barrel together with the gap for the downcomer add up to the smallest possible inner diameter of the vessel. At the cylindrical part of the vessel, the four circumferentially arranged inlets are positioned well above the four outlets, which are tilted by 45° in relation to the center-axis of the inlets (see Figure 9.1). The core barrel is suspended at the lower vessel top and

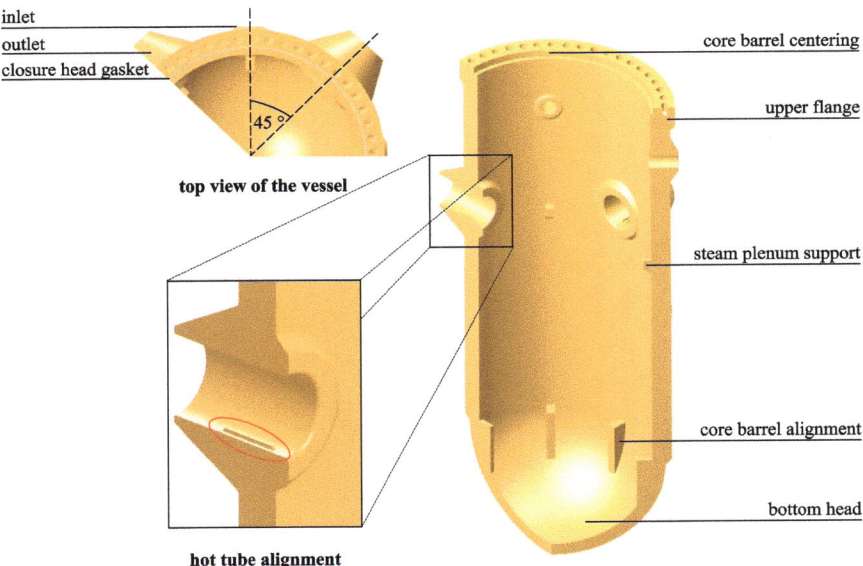

Figure 9.1: Lower reactor pressure vessel design for the three pass core design.

centered in azimuthal direction using four centering logs. The alignment in horizontal direction is realized using 8 support logs, located inside the bottom spherical section of the vessel. The core barrel sits, together with the control rod guide tubes, on a ledge machined from the RPV flange and is preloaded with a spring element. The lower vessel is braced with the closure head flange using reduced shank bolts and nuts (Figure 9.2). Two o-ring seals ensure leak tightness between the closure head and the lower reactor pressure vessel. The redundant design with two seals minimizes the possibility of leakage

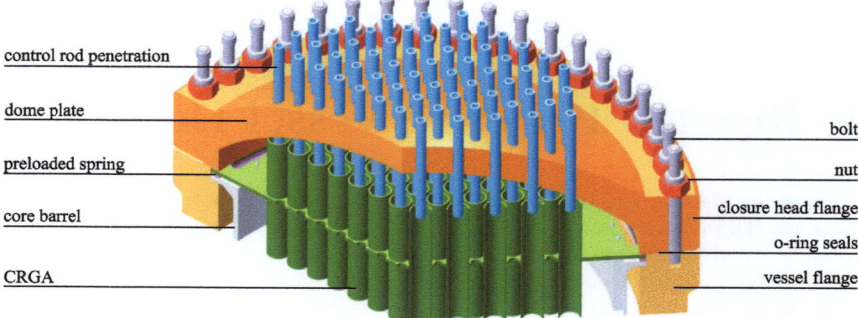

Figure 9.2: Closure head of the RPV with studs and o-ring seals for the three pass core design.

to the surroundings if one of them fails. Another design challenge results from the high temperature drop of over 220 °C for the hot pipes, which will have supercritical steam with a temperature of 500 °C in the inside of the tube, and the feed water coolant at 280 °C inlet temperature at the outside.

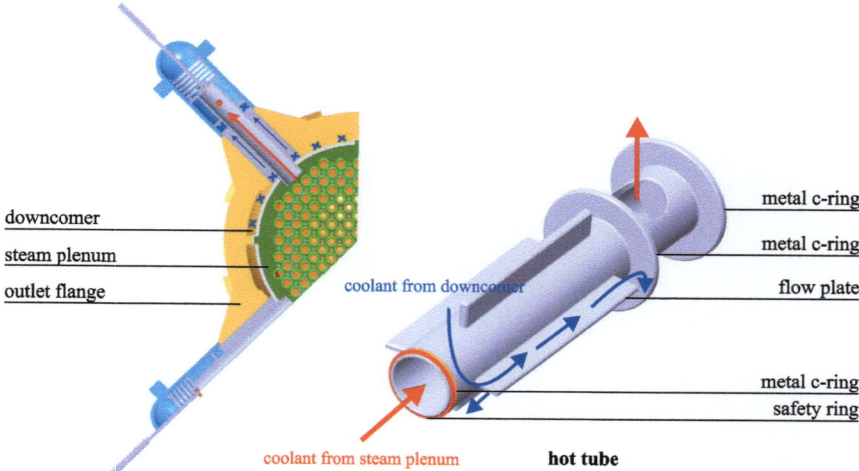

Figure 9.3: Cross section of the RPV at the level of the outlet flange (left side); schematic of the thermal sleeve for the hot tube with indicated flow path (right side) for the three pass core design.

The design for the thermal sleeve can be seen in Figure 9.3; on the left side a cross section of the RPV with internals at the level of the outlet flange is shown. The flow path of the supercritical water from the steam plenum to the outlet is marked in red, the flow path of the coolant from the downcomer, which forms the thermal sleeve around the outlet pipe, is blue. On the right side in Figure 9.3, one finned hot tube with flow paths and C-ring seal is illustrated. The horizontal flow plate forces part of the downward flowing downcomer water to take the marked path around it before the water flows back into the downcomer gap.

With this design a thermal contact of the hot pipes and the outlet nozzles is prevented. Therefore, only minimal thermal stresses and thermal deformations of the RPV are likely to occur.

10 Dimensioning of Critical Components

Boundary conditions, which have been summarized in the 5th framework program by Squarer et al. [83], are used for the dimensioning of the components. The proposed characteristics are modified by applying an increased safety factor of 115 % for the design pressure. In this case, for an operating pressure of 25 MPa a value of 28.75 MPa is used, giving a higher safety margin. The results presented here are all refering to the three pass core RPV design. Those components experience the highest loads, due to the bigger size and diameter of the core and therefore are the most critical concerning occuring maximum stress intensities and corresponding allowable material limits. The other two variants have also been analysed but will not be presented in detail in the scope of this thesis.

10.1 Candidate Materials for the Components

The reactor design includes the reactor pressure vessel with closure head and the internals with the core barrel with lower core plate and attached lower mixing plenum, the steel reflector, the steam plenum with adjustable hot pipes and the CRGA. For the vessel and closure head, the use of conventional vessel materials such as 20 MnMoNi 5 5 (United States: SA 508) is possible, due to the application of the thermal sleeve for the outlet making it possible to maintain a design temperature (T_D) of 350 °C for the inner wall of the RPV. Other important criteria for the material choice beside the design pressure and temperature include irradiation resistance and corrosion resistance for supercritical conditions. The boundary conditions for all reactor components are summarized in Table 10.1. The design pressure (p_D) of 28.75 MPa as a material criterium is only important for the vessel and closure head, since the other components are all located inside the vessel.

Component	T_D [°C]	p_D [MPa]	Neutron embrittlement
Lower vessel	350	28.75	very low
Closure head	350	28.75	very low
RPV bolts	300	-	very low
Core barrel	350	-	low
Core support plate	350	-	moderate
CRGA	350	-	very low
Steel reflector	400	-	high
Steam plenum	500	-	moderate
Hot tube	500	-	very low

Table 10.1: Boundary conditions for the material selection of the reactor components.

Candidate materials for those components are presently being investigated by several groups (Ehrlich et al. (2003, [22]), Was and Allen (2005, [95])). A main criterium for the material selection in the periphery of the core is the occurring neutron irradiation. The intense bombardment with neutrons creates dislocations in the materials, leading to radiation induced embrittlement, and to swelling in some cases. Generally it can be stated that metals exhibit an increasing tensile strength, hardness, and embrittlement with increasing neutron fluence. Additionally, increasing oxidation can be observed due to the corrosive nature of supercritical water in combination with the given temperatures. The selected materials have to show a high resistance against corrosion and a low susceptibility for stress corrosion cracking.

Existing materials for the application in supercritical fossil fired power plants can be used for the components with very low and low neutron irradiation due to comparable boundary conditions with temperatures of up to 600 °C and operation pressures up to 28.75 MPa. According to Ehrlich et al. [22] and Sridharan et al. [84] candidate components are the core barrel, the control rod guide assembly, and the hot tubes of the outlets. Those materials can be classified in three different groups, which will be discussed below.

The first group incorporates ferritic-martensitic steels like HCM12A, P 91, P 92, T 122 and 9 Cr ODS with chromium fractions of 9 to 12 % and additions of Mo, Nb, V, W and Cu depending on the alloy. Due to their high strength and high resistance against stress corrosion cracking, they are suitable for the core components (Allen et al. [2]). Another advantage is the operative experience due to the extensive use of those materials in fossil fired power plants with supercritical conditions. One drawback are the occurring high corrosion rates in water with high temperatures. To minimize this effect, additional chromium can be added to the alloy, with fractions as high as 12.2 % for T 122. W and Mo are added to enhance the creep resistance for high temperatures, one example is the ferritic martensitic steel HCM12A. Another method to increase the high temperature strength and therefore the creep resistance is the sintering of the base material with non-soluble, dispersive oxides. These materials are termed ODS-alloys for Oxide Dispersion Strenghtened (i.e. 9 Cr ODS).

The second group comprises austenitic steels, i.e. SS 316L, Incoloy 800, 1.4910, or TP 347 HFG. They feature a lower oxidation rate for supercritical water compared to the ferritic-martensitic steels (one order of magnitude lower) for temperatures between 400 and 550 °C [3] and are applicable for temperatures up to 650 °C [22]. Allen et al. [2] showed in a series of tests that local corrosion (pitting) takes place for temperatures around 300 °C and crevice corrosion occurs for temperatures around 500 °C. Only SS 316 L has been used for core components in actual BWR and PWR so far with data for long term behavior being available.

Nickel-alloys, like Inconel 625 and Inconel 690, are part of the third group for application temperatures over 650 °C. They feature further improved high-temperature strength and a higher oxidation resistance compared to the other two groups. Only preliminary results are available for those sophisticated materials, with research still going on to investigate long term behavior. Additionally, an increased neutron-absorption of up to 14 % can be observed due to the high fraction of Nickel compared to 10 % for the austenitic steels [22].

Using the boundary conditions from Table 10.1, a first material selection for the components can be executed. The core support plate, the steam plenum, and the steel reflector can be assigned to the second material group. For the first two, a material has to be selected with low neutron embrittlement and applicability for operation temperatures of 350 and 550 °C, respectively. Austenitic materials, in particular SS 316 L, where data for long term behavior is available, are the best choice in this case. The steel reflector is used to reduce the neutron flux escaping the core to optimize the utilization of the nuclear fuel. Another task is the protection of the RPV against fast-neutron fluence-induced aging and embrittlement. No data on corrosion behavior for supercritical conditions is available, since applications of this reflector are only known for LWR using subcritical water. Therefore, the material X 2 CrNi 19 10 is chosen for the steel reflector which has been applied for the newest design of the PWR (EPR, [31]). The high chromium fraction provides a quite good corrosion behavior for this austenitic steel, but the real behavior has to be certified by material tests.

The core barrel and the CRGA are exposed to quite low neutron fluence with neutron embrittlement playing a minor role, additionally the design temperature of the material is only around 350 °C. In this case ferritic-martensitic materials are the best choice. A representative material is P 91 (X 10 CrMoVNb 9-1) with a chromium fraction of 9 %, which is also quite common for supercritical fossil fired power plants. It is also used for the outlet tubes, which have to be welded to the ferritic material of the RPV flanges. Due to the higher temperatures of the supercritical steam with 500 °C, materials of the first group cannot be used for the hot tubes. The same austenitic steel as for the steam plenum is used in this case being more suitable for the high temperatures. For the other components spring element, hold-down spring, and RPV bolt, typical materials of LWR are chosen due to the similar boundary conditions.

This first material selection for the components has been made to obtain material strength characteristics for the dimensioning. The chosen material for each component is listed in Table 10.2. This material selection is entirely tentative and may change when data on materials being tested against oxidation, stress corrosion cracking, etc. become available.

Material	Component
X 10 CrMoVNb 9-1 (P 91)	core barrel, CRGA, outlet tubes
SS 316 L (N)	steam plenum, core support plate, lower mixing plenum
X 2 CrNi 19 10	steel reflector
Inconel 718	spring element, hold-down spring FA
24 CrMo 5	RPV bolt, nut

Table 10.2: Material selection for the reactor components.

10.2 Mechanical Analysis

The dimensioning of the RPV and closure head as well as the design calculation for the studs, nuts and o-ring seals is performed using the safety standards. The boiler equation serves as the basis to calculate the different wall thicknesses of the cylindrical and the spherical parts of the vessel. Table 10.1 on page 85 gives an overview of some important parameters for the RPV and the closure head. Due to forging limits, a general constraint for the design of the reactor pressure vessel is its maximum wall thickness. To minimize the wall thickness, which is exposed to the high design pressure of 28.75 MPa, a preliminary mechanical analysis is performed in accordance with the safety standards of the nuclear safety standards commission (KTA) in Germany [52]. The dimensions of the RPV with closure head, inlets and outlets for the three pass core are depicted in Figure 10.1. Note that the thickness of the upper pressure vessel head depends on the number and type of penetrations, which have to be defined by core analysis. The design presented here gives an example for one possible arrangement of the control rod drive mechanism guides penetrating the spherical part of the closure head. According to Boungiorno and MacDonald [12], it is feasible to manufacture a vessel with the mentioned dimensions. Even a 4.2 m high forging around the core, to prevent circumferential welds from the high fluence beltline section, seems possible.

A second stress analysis is performed for the core internals with high working load. This is the case for the flange of the core barrel and the flange of the core support plate, carrying the weight of the core internals and the support brackets of the RPV, where the steam plenum is mounted. Additionally, the studs and nuts of the closure head have been dimensioned using the

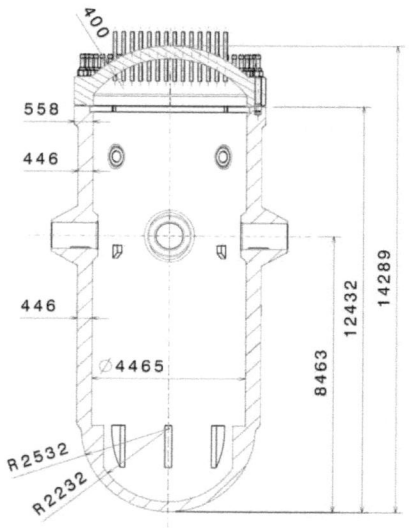

Figure 10.1: Dimensions of the RPV and closure head in mm for the three pass core design.

safety standards. The 40 studs are chosen as reduced shank bolts according to the German DIN standard 2510 for pressure vessel bolts with a diameter of 162 mm for the critical cross section. A spring element, which is positioned between the flange of the core barrel and the closure head is dimensioned in a way, that the gaskets of the closure head are not damaged by the pre-load force of the reduced shank bolts of the RPV.

Three critical cross sections are defined for the core barrel flange and the core support plate flange using the safety standards. For the calculations, the weight of the components has to be determined using the density of the suggested materials. Table 10.3 summarizes the required values for the analysis. The weight load for the core barrel flange amounts to 455.9 t, including the core, the core support plate, the lower plenum, the steel reflector, the CRGA and the core barrel itself, resulting in a wall thickness of 0.06 m for the cylindrical part, and a flange height of 0.075 m.

The same approach is used for the core support plate flange, which has been dimensioned with a thickness of 0.06 m. The four support brackets for the steam plenum are dimensioned in such a way that each bracket is able to carry the complete weight of the steam plenum. This is the worst case scenario, which can occur during the assembling, when the steam plenum is lowered on the support brackets. Therefore, each bracket is

10.2 Mechanical Analysis

designed to bear a maximum load of 8.6 t on this minimum bearing surface. A summary of the occurring maximum stress intensities for the investigated components is given in Table 10.4.

Component	Weight [t]
Lower vessel with outlet tube	671
Closure head with bolts and nuts	164
Core with fuel assembly clusters	210
Core barrel	61.1
Steel reflector	104
CRGA	54
Core support plate	19
Lower plenum	7.8
Steam plenum	8.6

Table 10.3: Weight of the different reactor components in t.

Due to the highly varying temperatures of the coolant and the materials inside the vessel, thermal expansions between the different components have to be considered. For example, the core barrel with a height of approx. 9.7 m will expand 0.05 m in the vertical direction when heated from ambient conditions to a coolant temperature of max. 350 °C. The thermal expansions of critical components like the core barrel, steel reflector and the CRGA are calculated and set in relation to the other affected internals using the equations from section 2.5.2 on page 23 with the corresponding thermal efficiency coefficient of the material. The different expansions of the components are also summarized in Table 10.4.

Component	Max. stress [MPa]	Expansion [m]
Lower vessel	210	-
RPV bolt	105	-
Core barrel	241	0.05
Steel reflector	-	0.03
CRGA	85	0.02
Core support plate	232	-
Steam plenum support	40	-

Table 10.4: Maximum stress intensity of the investigated components with calculated thermal expansions.

Using the expansions of the single component from Table 10.4, the thermal expansions in relation to the other afflicted internals are calculated to determine corresponding safety margins which have been implemented into the design.

11 Design Verification Using Finite Elements Methods

Due to different coolant temperatures and thus material temperatures, the reactor pressure vessel with its feedwater and steam flanges experiences large thermal deformations. To verify the concept of the pressure vessel with the outlet flanges, a combined mechanical and thermal finite elements analysis (FEM) with the software program ANSYS is performed for the different components. The results are evaluated using the KTA standard 3201.2 [52] which is concerned with the components of the reactor coolant pressure boundary of LWR, including the design and analysis of the pressure vessel, the feedwater inlets, the main steam outlet nozzles and the closure head. The results of the analysis for the three pass core RPV and steam plenum design will be presented by Fischer et al. [27].

The steam plenum on top of the core is the other component with high thermal stresses and deformations. For the three pass core it is built from an inner part, in which coolant of 390 °C at the outlet of the first heat up step is mixed before entering the superheater, and an outer part, which collects and directs the live steam of 500 °C to the steam flanges. The steam plenum is surrounded by feed water of 280 °C. These high temperature differences cause thermal deformations and stresses which are analyzed using finite elements in accordance with the KTA-guidelines for reactor pressure vessel internals [50]. The occurring deformations have to stay inside a prescribed limit for the sealings to be tight.

The dimensioning criteria for both investigations are based on the design loading level (level 0) and take into account the loadings and service limits of level A as far as they concern dimensioning. The evaluation and superposition of stresses are carried out for each load case where the stresses acting in the same direction are added separately or for different stress categories jointly. The allowable values for stress intensities and equivalent stress ranges for the linear-elastic analysis of the mechanical behavior for the analyzed loading levels 0 and A are determined according to section 2.5, page 21. For the analysis of the peak stresses, transient loads with an estimate of 5000 cycles are considered for all components to determine the allowable half stress intensity range S_a as described in section 2.5.

11.1 Geometry and Numerical Model

The mesh for the geometry of the pressure vessel, its components and the steam plenum for the three pass core design has to represent the geometrical features to an adequate resolution, so that the discretization error is small enough to satisfy the recommended accuracy. For the complex geometries, structured meshes are very difficult to apply, therefore unstructured meshes are preferred since they are the most flexible and adaptive to an arbitrary solution domain boundary. The disadvantage lies in the irregularity of the element and node structure which slows down the solution process compared to the structured meshes.

The geometry of the reactor pressure vessel, its components and the steam plenum is generated with the CAD package of CATIA and imported into ANSYS WORKBENCH. To ease the computational effort, the symmetric arrangement of the reactor pressure vessel flanges is used. To represent the occurring loads, stresses and deformations, it is sufficient to model a segment of one-eighth or 45° of the axis-symmetric geometry. The first symmetry plane cuts through the middle of the outlet flange and the outlet pipe, while the second plane cuts through the middle of the inlet flange. Appropriate boundary conditions have to be considered for the segment cutting planes, which will be discussed in the next section.

The same approach as for the reactor pressure vessel is used for the steam plenum. A segment of one-eighth or 45° of the axis-symmetric geometry with one segment plane cutting through the outlet of the steam plenum and the other one through the middle of the guide strip is modeled. The resulting computational domain can be seen in Figure 11.3 and 11.4, respectively.

For the locally refined meshes the following evaluation method is applied for post-processing. The calculated von-Mises stress distribution is used to determine the critical zones where a cross section for the evaluation is defined. In the case of the upper flange (see Figure 11.1) the appropriate evaluation surface already corresponds to the symmetry plane, otherwise it is created in a way to capture the stress distribution throughout the critical zone.

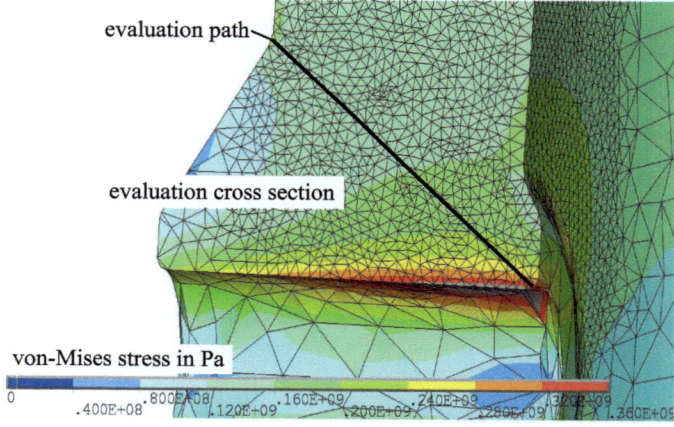

Figure 11.1: Evaluation cross section with applied evaluation path for the refinement of the upper part of the inlet flange.

An appropriate number of evaluation paths are mapped on the surface to represent the existing stress distribution. The evaluation path is used to separate stresses into the individual components for the analysis. The calculated and reviewed linearized stresses are necessary for the evaluation of the primary, secondary and peak stresses which are defined for the two loading levels 0 and A mentioned in section 2.5.1, on page 21. Due to

11.1 Geometry and Numerical Model

the unstructured character of the grid, it is not possible to go along a line of nodes for the evaluation, the resulting stresses have to be interpolated along the path. The number of 30 path points is set high enough to resolve the stress pattern.

The graph in Figure 11.2 shows the resulting linearized stress distribution for the upper part of the inlet flange for the illustrated evaluation path in Figure 11.1 versus the path distance range. The orange graph resembles the linearized membrane stress distribution which is a primary stress, while the red graph symbolizes the distribution of the combined membrane, bending and thermal stresses. The blue graph shows the distribution of the linearized stresses of the red graph in combination with peak stresses. To analyze the critical zone, the occurring maximum stress for each graph is compared with the allowable material limit.

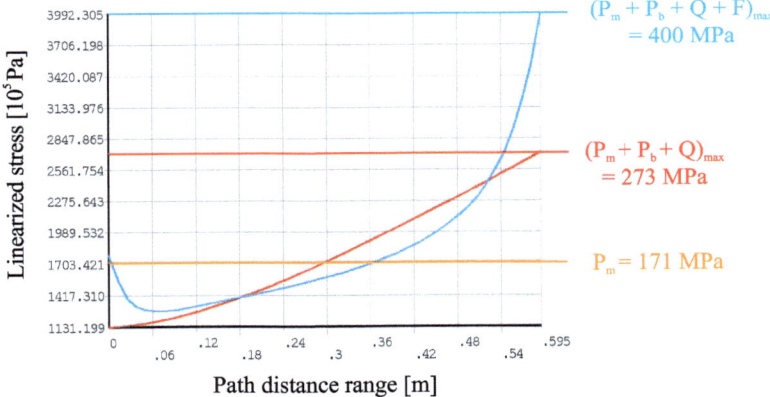

Figure 11.2: Evaluation graph for the linearized stress distribution (ordinate) versus path distance range (abscissa).

To evaluate the calculation results for their plausibility in physical respect, all critical zones with peak stresses are investigated using different local refinements to observe the behavior of the von Mises stress distribution. In the case of physical discontinuities, those stresses increase with finer meshes and concentrate on a small number of elements leading to singularities in those zones. This effect can be attributed to the simplified geometry in combination with the applied linear material behavior where plastification effects are not modeled. This material plastification is the major process to dissipate occurring peak stresses inside of notches and sharp edges in the geometry. Those peak stresses with the resulting singularities can be omitted by reshaping and optimizing the geometry. The investigated structure is idealized, so that neither local nor global singularities of the stiffness matrix occur. Otherwise, they are observed and evaluated if they lead to an adulteration of the physical behavior. Another singularity effect is attributed to wrongly placed boundary conditions and degrees of freedom. The numerical code is able to detect such inconsistencies and allows the user to improve the conditions.

11.2 Boundary Conditions

For the first analysis, the reactor pressure vessel, the outlet flanges, the hot pipe and the outlet pipe are modeled. The internals are included as a mechanical boundary condition for the reactor pressure vessel flange. The modeled 45° segment of the RPV with outlet can be seen in Figure 11.3. The first symmetry plane is positioned in the middle section of the outlet flange and the outlet pipe, while the second symmetry plane cuts through the middle section of the inlet. These two different supports for the vessel allow expansions and deformations in radial direction and are defined as frictionless supports. The fixation in vertical direction is realized with a frictionless support in the circumference below the outlet flanges. The RPV features eight of these support brackets resting on the concrete basement which surrounds the vessel.

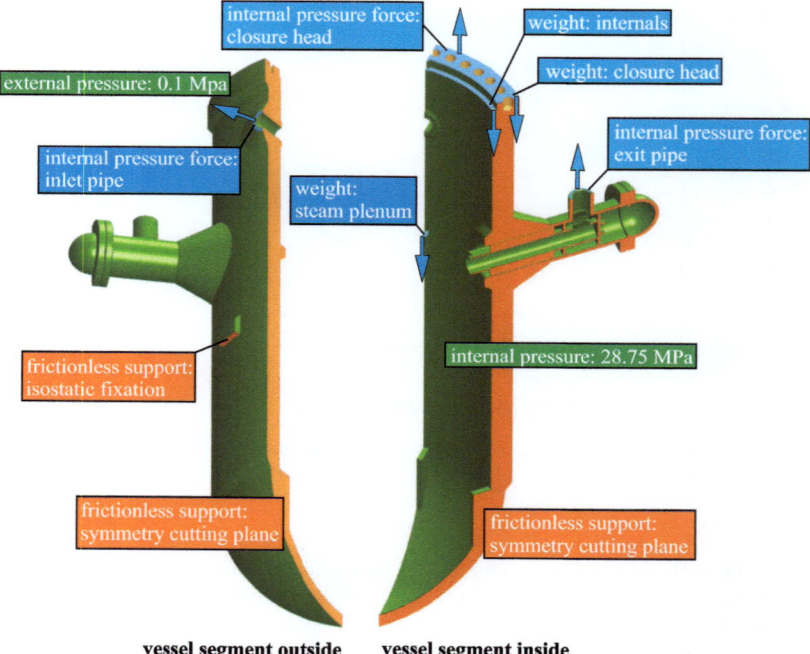

Figure 11.3: Mechanical boundaries for the 45° segment of the RPV with steam outlet flange including the external loads resulting from the internals.

Mechanical loads for loading level 0 include the design pressure of 28.75 MPa for the inside of the vessel, a pressure of 1 MPa applied to the outside surface of the vessel and the design temperature. Additional mechanical loads comprise the weight of the core internals and the closure head, but also resulting forces on the inlet, outlet and RPV flange area due to the internal pressure in the pipes and vessel.

For loading level A all secondary stresses and peak stresses are considered in addition

11.2 Boundary Conditions

to the primary stresses for level 0, in this case these are the thermal stresses caused by the different temperatures of the components and parts of the RPV. To model the convections between the different parts for the temperature boundary condition, the heat transfer coefficient has to be determined for each case. The corresponding heat transfer correlations for the occurring flow regime as described in section 2.6 on page 23 are applied. Additionally, the mass flows for the downcomer and the thermal sleeve are estimated using pressure drop calculations as described in Guelton and Fischer (2006, [37]). The resulting temperatures and heat transfer coefficients for the calculation of the temperature field in ANSYS are given in Table 11.1.

Component	Temperature [°C]	α [Wm^{-2} °C^{-1}]
RPV		
Inner wall, flanges	280	3020
Outer wall; outlet pipe, outside	150	10
Hot pipe, outside	280	2020
Hot pipe, inside; outlet pipe, inside	500	10500
Steam plenum		
Top cover; top FA openings	280	500
Peripheral cover; guide strip	290	2000
Bottom cover; bottom FA openings	300	500
Inner wall, eva /sh 1	390	13288
Inner wall, sh 2	500	3235
Connection tubes, inside	280	4160
Connection tubes, outside, eva	390	13938
Connection tubes, outside, sh 1	390	12637
Connection tubes, outside, sh 2	500	3235
Separation wall, sh 1	390	12637
Separation wall, sh 2	500	3235

Table 11.1: Thermal boundary conditions temperature and heat transfer coefficient α for the RPV with outlet tube and the steam plenum with evaporator (eva), superheater 1 (sh 1) and superheater 2 (sh 2) region.

The second analysis is performed for the steam plenum with connection tubes, coolant exits, and separation wall. The modeled 45° segment of the steam plenum with all components can be seen in Figure 11.4. The first radial symmetry plane is positioned in the middle section of the coolant exit, while the second symmetry plane cuts through the middle section of the guide strip.

Mechanical loads for loading level 0 include the external pressure of 28.75 MPa for the outside shell of the steam plenum and the inside of the connection tubes, an internal pressure of 28.66 MPa for the upper mixing plenum and an internal pressure of 27.35 MPa for the exit zone.

The heat transfer coefficients for loading level A are determined for each case as described for the first analysis using corresponding heat transfer correlations from section 2.6 for the occurring flow regime. The thermal boundaries include the temperatures and heat transfer coefficients given in Table 11.1 for the different surfaces.

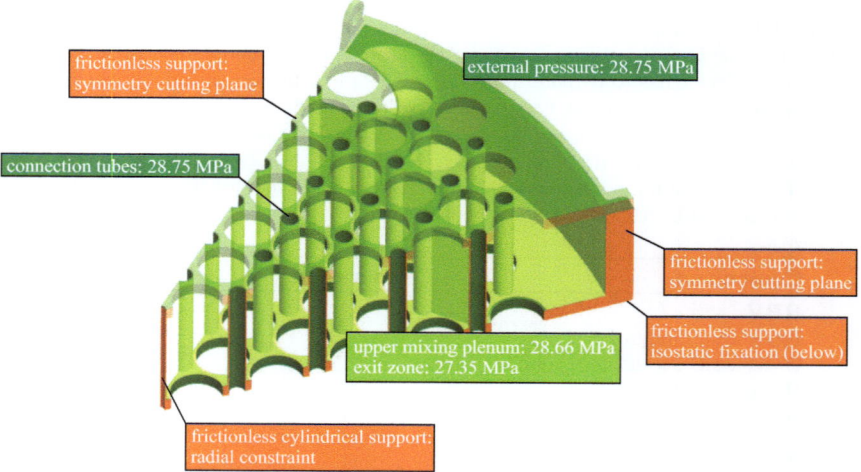

Figure 11.4: Mechanical boundaries for the 45° segment of the steam plenum with separation wall and connection tubes.

11.3 Temperature Distribution and Deformation Analysis

The modeled temperature field from ANSYS for the RPV and steam outlet flange is illustrated in Figure 11.5. It can be seen, that the high temperature of 500 °C for the inside of the hot pipe through the flow plate to the inner side of the outlet flange is constantly decreasing, so that the temperatures are around 280 °C in the zone of the outlet flange. The analysis states that the calculated mass flow for the thermal sleeve is sufficient to isolate the high temperatures of the hot pipes from the outlet flanges of the reactor pressure vessel. The upper vessel flange experiences thermal deformations due to the different temperatures for the inlet and outlet flange (see Figure 11.5) influencing the leak tightness of the two applied o-ring seals for the closure head. Especially the difference between the deformations above the outlet and the inlet is important for the evaluation. A sample path is applied along the sealing surface for the modeled geometry (see Figure 11.6, left side). The minimum displacement occurs above the outlet flange while the maximum is determined between the inlet and the outlet flange. This trend is not only attributed to the occurring temperature differences, where the maximum would be expected above the outlet flange, but also to structural discontinuities like the penetration in the shell leading to a smaller deformation above the outlet. On the right side of Figure 11.6, the evaluation graph for the sample path is shown with the total deformation (ordinate) versus path distance range (abscissa) yielding a small enough deformation difference of 0.016 mm to have no impact on the sealing of the closure head.

The modeled temperature field for the calculated thermal boundaries of the steam plenum are displayed in Figure 11.7. The exit zone and especially the exit nozzles ex-

11.3 Temperature Distribution and Deformation Analysis

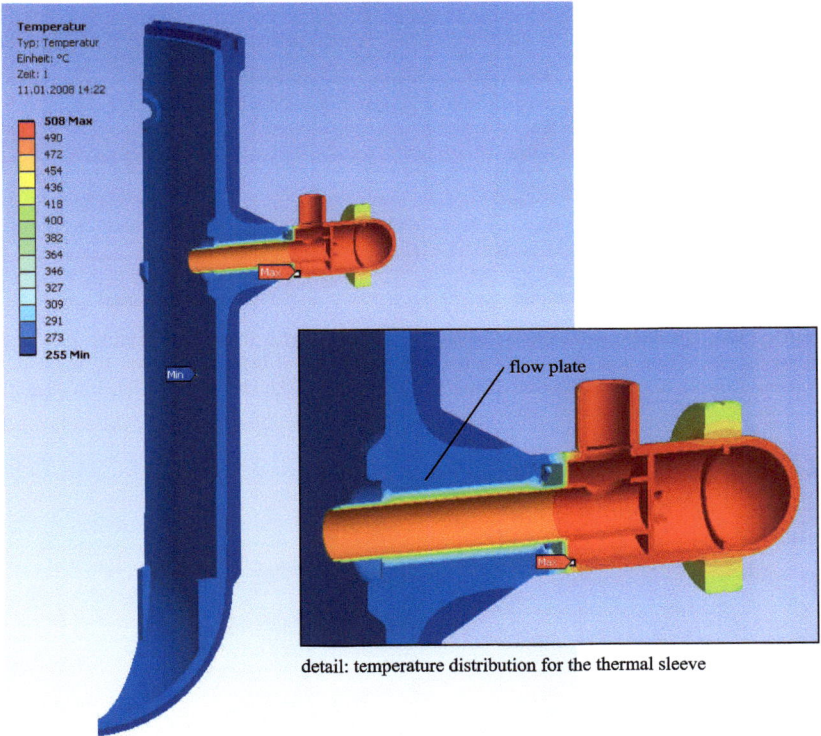

Figure 11.5: Calculated temperature distribution from ANSYS for the RPV and thermal sleeve at the outlet (detail on the right side) for the given thermal boundary conditions.

perience the highest temperature of 500 °C. Due to the temperature difference in the two chambers and the surrounding cold moderator water, the fuel assembly openings are exposed to a quite high temperature difference of up to 200 °C yielding considerable thermal deformations. The resulting deformations of the modeled segment can be seen in Figure 11.8. The maximum deformation occurs in the vicinity of the guide strip which can be attributed to the applied load from the weight of the steam plenum. To verify the leak tightness of each fuel assembly cluster opening for the top and bottom sealing, the deviation of each opening from the non-deformed structure is measured every 45° along the periphery of the opening, shown in Figure on the right side 11.8. Maximum deviations are experienced by the peripheral openings for both covers yielding values around 0.275 mm. Those deformations are still ensuring leak tightness between the fuel assembly cluster and the steam plenum for the applied C-ring seals [35]. For the exit nozzle, the deformations are quite homogeneous and do not affect the leak tightness between the hot pipe and the sealing surface.

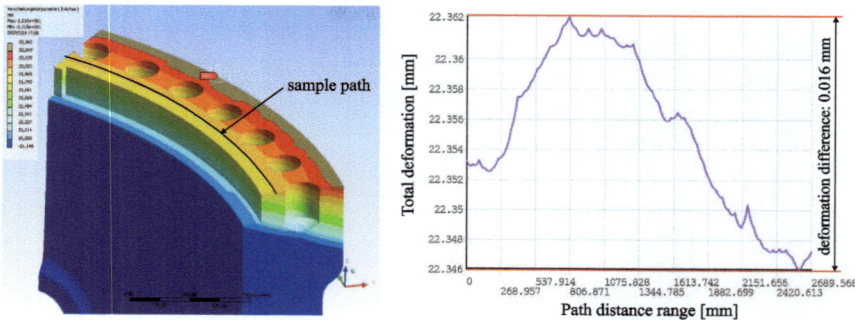

Figure 11.6: Displayed deformation of the vessel flange with illustrated sample path in the vicinity of the o-ring seal (left side, a 5 times magnified illustration is used to display the occurring deformations); evaluation graph for the sample path showing the total deformation (ordinate) versus path distance range (abscissa) (right side).

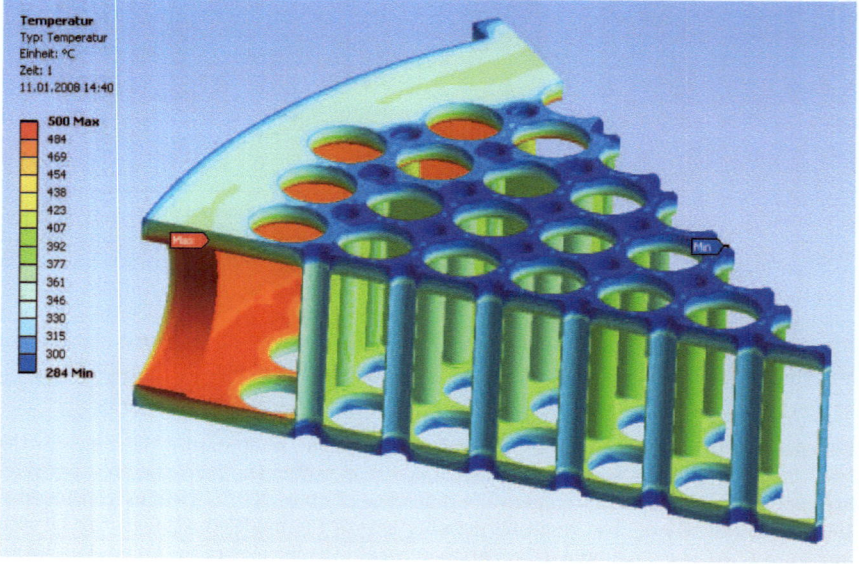

Figure 11.7: Temperature distribution in ANSYS for the modeled segment of the steam plenum using the calculated thermal boundary conditions.

11.4 Evaluation of Stress Intensities

The combined thermo-mechanical analysis yields the maximum observed stress intensity σ_{max} for the critical cross section for each component of the RPV. The linearized maximum primary (category I and II), secondary (category III) and peak (category IV)

11.4 Evaluation of Stress Intensities

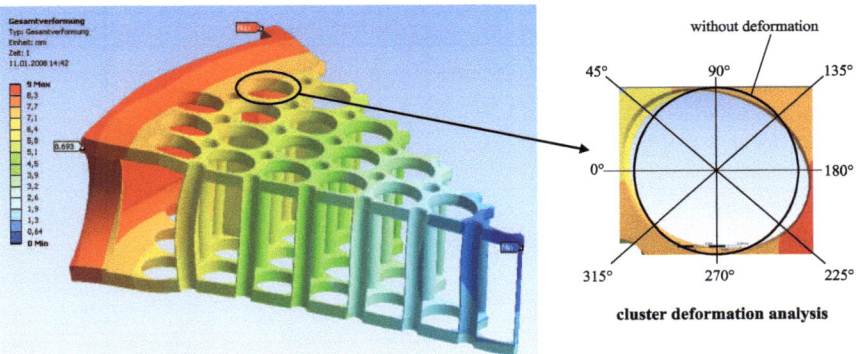

Figure 11.8: Resulting deformations in mm of the modeled segment of the steam plenum for the applied thermal and mechanical boundary conditions; example deformation analysis of a cluster opening using eight evaluation points on the periphery (right side). A 17 times magnified illustration is used to display the occurring deformations.

stresses for the critical cross sections as defined in section 2.5.1 on page 21, together with the corresponding allowable stress intensity σ_{al} for the investigated components vessel, outlet pipe, and hot pipe of the RPV using the corresponding material selection from section 10.1 can be seen in Table 11.2. For the analysis of the peak stresses, transient loads with an estimate of 5000 cycles are considered for all components to determine the allowable half stress intensity range S_a for ferritic and austenitic steels according to Appendix A.1.1 and Appendix A.1.2, respectively. The ratio between the maximum observed stress and the allowable stress for the different categories yields the material utilization in percent. It can be stated that all components stay within the prescribed material limits. The maximum stresses for the lower vessel occur for all four categories in the vicinity of the outlet flange. This can be attributed to the large penetration for the outlet in the cylindrical shell with its non-uniform stress distribution due to the superposition of tangential and longitudinal stress intensities. The occurring thermal stresses also have an impact on the magnitude of the stress intensity. The chosen design with the inner stiffening ring for the outlet flange helps to homogenize the stress distribution. The same phenomenon can be observed for the outlet pipe, where the maximum stress intensity occurs in the vicinity of the ovally shaped outlet. Since the utilization of the material is around 77 % at maximum, a supporting stiffening ring in this area is not necessary. The outlet pipe experiences quite high temperature differences over the cross section of the flow plates, resulting in maximum secondary and peak stresses for this component. The maximum of the primary stresses occcurs for the bottom flow plate, where the weight of the hot pipe rests on. Still, all occurring maximum stresses stay well inside the material limits, with utilization values around 47 % at maximum.

A separate analysis is performed to verify the wall thickness of the upper pressure vessel head with the several guides penetrating the spherical part of the closure head. The arrangement and number of penetrations is symmetric, therefore a segment of 45° is sufficient for the modeling. The von-Mises stress distribution for the closure head with the

Component	Stress category	σ_{max} [MPa]	σ_{al} [MPa]	Utilization [%]
Lower vessel				
Outlet flange	I	170	210	80.9
Outlet flange	II	282	315	89.5
Outlet flange	III	304	630	48.3
Outlet flange	IV	478	700	68.3
Outlet pipe				
Cylindrical shell	I	140	198	70.7
Cylindrical shell	II	165	297	55.6
Exit zone	III	275	594	46.3
Exit zone	IV	541	700	77.3
Hot pipe				
Flow plate	I	45	179	25.1
Flow plate	II	84	269	31.2
Flow plate	III	250	537	46.6
Flow plate	IV	541	700	41.3
Steam plenum				
Exit nozzle	I	148	198	74.7
Exit nozzle	II	224	297	75.4
Exit nozzle	III	585	594	98.5
Opening FA	IV	601	960	90.8

Table 11.2: Summary of the maximum observed stress σ_{max} with corresponding allowable stress limit σ_{al} and material utilization from the coupled thermo-mechanical analysis for the lower vessel, hot pipe, outlet pipe, and steam plenum.

illustrated penetrations for the guides can be seen in Figure 11.9. Inhomogeneities in the stress intensity distribution result from the openings due to the penetrating guides. Peak values have been identified in the vicinity of the penetrations with the highest values of up to 300 MPa close to the center of the spherical shell. These values are not critical and are still below the allowable material limit. The linearized maximum primary (category I and II), secondary (category III) and peak (category IV) stresses for the critical cross sections, together with the corresponding allowable stress intensity σ_{al} for the investigated steam plenum can be seen in Table 11.2. The observed maximum stress intensity σ_{max} for loading level 0 can be found for the exit nozzle at the height of the opening midpoint. This maximum results from the superposition of the tangential and longitudinal stress for the opening in the cylindrical shell. Those stress intensities are still acceptable and deformations are kept low due to the applied conical stiffening for the nozzle. For the investigation of the peak stresses, the maximum can be observed in the vicinity of the peripheral bottom fuel assembly openings, where the high temperature differences between the inside and outside of the plate dominate the peak stress distribution. All parts of the steam plenum stay within the given material limits, with utilization values varying between 74.7 and 98.5 %.

In summary, it can be stated that the optimized design for the RPV of the HPLWR with a three pass core design has been verified with the finite element analysis, including

11.4 Evaluation of Stress Intensities

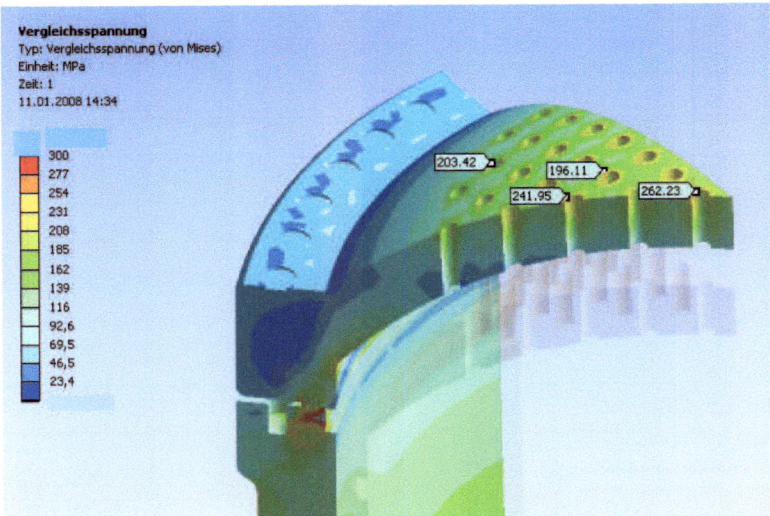

Figure 11.9: Von Mises stress distribution for the RPV closure head with shown peak von Mises stress values for the several guide penetrations.

mechanical and thermal loads. With an inner vessel diameter of 4.47 m and a maximum wall thickness of 0.45 m in the cylindrical part, the pressure vessel can still be forged. The thermal analysis states that the calculated mass flow for the thermal sleeve is sufficient to isolate the high temperatures of the outlet pipes from the outlet flanges of the reactor pressure vessel. Additionally, the thermal deformations due to the different temperatures for the inlet and outlet flange are small enough to ensure leak-tightness of the two o-ring seals for the closure head.

The design of the steam plenum also has been verified using a coupled thermal-structural analysis. The occurring deformations and stresses are acceptable, due to the decoupling of the steam plenum nozzles from the guide strips and the conical stiffening of the nozzles. Leak-tightness of the steam plenum is assured for all fuel assembly openings and exit nozzles with maximum deviations of 0.275 mm from a preferable ring shape. C-rings, featuring the size of the openings, have tolerances in the order of 0.3 mm.

12 Fluidic Optimization of the Backflow Limiter

This chapter describes the application of a backflow limiter for the RPV inlet of the HPLWR in order to reduce the mass flow rate in case of a LOCA for a feedwater line break. Starting with a simple design of a vortex diode, several steps are executed to enhance the performance of the diode and adapt it to this application. The backflow limiter and its optimization has been discussed by Fischer et al. [25] and will be presented in [26].

The operating principle of the simple vortex diode is illustrated in Figure 12.1. In the reverse flow direction (A) the fluid enters through the tangential port and forms a vortex in the vortex chamber. This dynamic head is destroyed when the flow discharges through the axial port giving the high flow resistance. For regular operation condition (B), the flow enters through the axial port without forming a vortex and leaves the chamber through the tangential port resulting in a small pressure loss only.

For the adaptation to the RPV inlet feedwater line, the task is to optimize the component as much as possible in regular operation condition reducing the pressure loss to a minimum, without giving up too much of the performance i.e. high resistance in reverse flow condition. The diminished pressure loss for regular operation condition directly reduces the costs for the feedwater pumps. Therefore, several adaptations for the simple vortex diode are discussed.

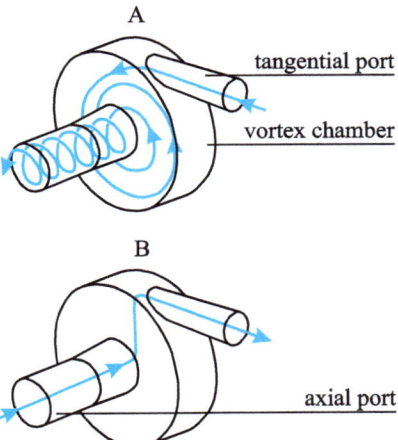

Figure 12.1: Flow path for reverse flow condition (A) and regular operation condition (B) for the simple vortex diode, according to Baker [5].

One major part of the pressure loss in regular operation condition results from the diffusion of the flow into the big volume of the vortex chamber. Furthermore, the diffusion is not directed to the tangential port, but radially into the whole chamber. Therefore, the flow has to bend at the outer periphery of the vortex chamber to flow into the tangential port. Flow separation in this radial diffuser can be avoided by imposing a low swirl on the flow inside the axial port. This can be achieved with inlet swirler vanes which are radially arranged inside the axial port. Another benefit of this design change is the additional resistance in the reverse flow condition, when the vanes enhance the pressure loss by destroying the vortex inside the axial port. Sidhu et al. [79] investigated a vortex diode, where multiple tangential ports are used. The pressure losses in regular operation condition are considerably reduced due to the higher number and symmetric arrangement of the tangential ports, increasing the flow cross section of the vortex chamber. They can even be replaced by circumferentially arranged exit vanes. For the reverse flow condition, the higher number of ports helps to maximize the intensity of the vortex flow.

Applying an optimized diffuser section to the tangential ports and to the axial port for regular operation condition gives another reduction in pressure loss, due to the pressure recovery effect of the diffuser (Zobel [101]). In reverse direction the diffusers act as nozzles and help to maximize the intensity of the vortex flow.

To reduce the bending losses at the transition from the axial port to the vortex chamber, the sharp edge can be rounded off, but this also decreases the resistance in reverse flow direction. Additionally, a center body in the vortex chamber helps to prevent separation of the flow. Moller [59] found that the pressure losses are minimum if the contour for the center body is approximated with a constant-area bend.

A general guideline for the design of turbo engines from Sigloch [81] is used to determine the shape and number of inlet and exit swirler vanes for the adapted vortex diode. The inlet swirler vanes are designed like guide vanes of an axial water turbine. A simple plate vane profile is chosen with a radially increasing cross section to account for the smaller circumferential velocity with increasing diameter. The number of vanes depends on the diameter of the flow channel and the induced swirl. Design criteria for guide vanes in radial pumps are applied, which recommend that the smallest cross section should be square to minimize pressure losses. Additionally, the inlet and exit area are limited by the RPV size and the number of exit vanes is influenced by design criteria like the inlet vanes. Preferably, the number of exit vanes should be multiples of the inlet vanes to ease CFD analyses. Furthermore, the angle of the diffuser should not exceed 8° to avoid recirculation zones. An evaluation of the considered optimization procedures for the simple vortex diode is given in Table 12.1, where the (+) symbolizes a positive effect for the pressure loss reduction in normal operation condition (B) or reverse flow direction (A) and (-) a negative effect, respectively.

Optimization procedure	B	A
Impressed preswirl (i.e. vanes in axial port)	+	+
Multiple tangential ports	+	+
Smooth walls (negligible wall roughness)	+	+
Diffuser section for the tangential port	+	+
Diffuser section for the axial port	+	+
Center body in the vortex chamber	+	
Smoothed transition axial port ⇔ chamber	+	-
Optimized relation swirl angle ⇔ outflow angle	+	-
Increased diameter and height of the chamber		+

Table 12.1: Evaluation of the considered optimization procedures for the simple vortex diode; (+) corresponds to a positive effect for the pressure loss reduction in normal operation condition (B) and (-) a negative effect.

These changes are made to improve the regular operation condition using the simple vortex diode as a reference. To improve the overall performance, some additional flow resistances for the reverse direction are applied, which do not considerably affect the optimized resistance in regular operation condition. For larger scale vortex diodes, i.e. a larger outlet diameter and height of the exit vanes, the performance of the investigated diode will improve due to the stronger vortex in reverse direction. Therefore, it is desirable to increase the diameter and the height as much as possible. In this application, the limiting factor is the space for the vortex chamber inside the RPV. It is defined by the space between the core barrel and the inner wall of the RPV.

Figure 12.2 shows a section of the RPV of the three pass core design at the height of the feed water inlet with the implemented backflow limiter. The inner diameter of the annulus that can be used for the backflow limiter has a diameter of 4.23 m. The outer diameter is given with 4.465 m, which represents the inner diameter of the RPV. As a result, the maximum possible exit diameter of the backflow limiter is 0.95 m and the maximum possible height of the exit vanes is 0.02 m. The number of exit vanes has been chosen such that the flow cross section has a quadratic shape at the smallest cross section of the exit vanes to minimize flow losses. The inlet tube for the backflow limiter features a fixed diameter of 0.2 m.

This first design is optimized using 1D flow analyses to determine the geometrical shape of the backflow limiter and the swirler angles. In an iterative process, the design is further optimized by determining the pressure loss for specific sections of the diode using 3D CFD analyses and minimizing them by changing the geometry parameters in the 1D analysis. As a last step, the swirler angles are varied to find an optimum for minimal pressure losses in the regular operation condition. Finally, parametric studies determine the performance factor for changing mass flow rates to receive the characteristic of the backflow limiter.

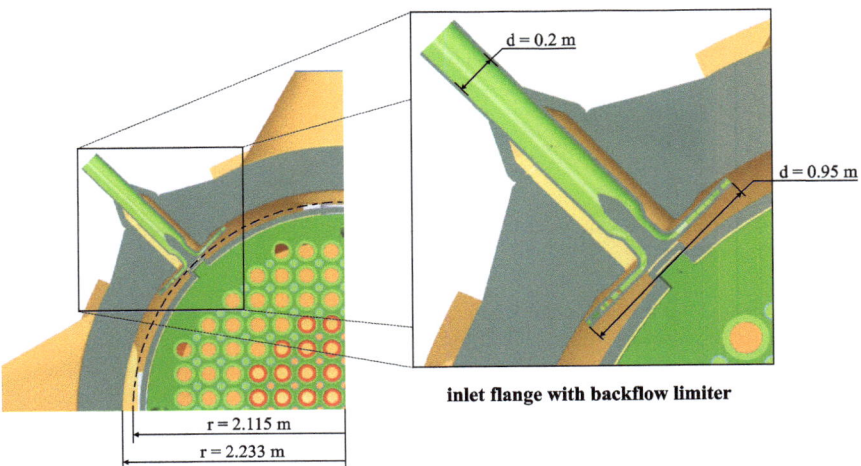

Figure 12.2: Quarter section of the RPV for a three pass core (left side) with implemented backflow limiter in the inlet flange (right side).

12.1 Design of the Backflow Limiter

The resulting design for the backflow limiter, considering the presented suggestions, can be seen in Figure 12.3. In extension of the center body of the vortex chamber, a central carrier for the vanes has been implemented which ends with a conical inlet piece. This design has been chosen to strengthen the mechanical stiffness. The conical inlet piece accelerates the flow in regular flow condition while it gives additional resistance in the reverse flow direction. The axial nozzle behind the inlet swirler vanes (shown in blue in Figure 12.3) causes additional resistance in reverse flow direction, while the additional losses for the regular operation condition are negligible (Moller [59]). The design features 10 inlet swirler vanes and 30 exit swirler vanes. The outlet angle of the optimized inlet swirler vanes is 10°; the exit swirler vanes have an outlet angle of 60° for regular operation condition.

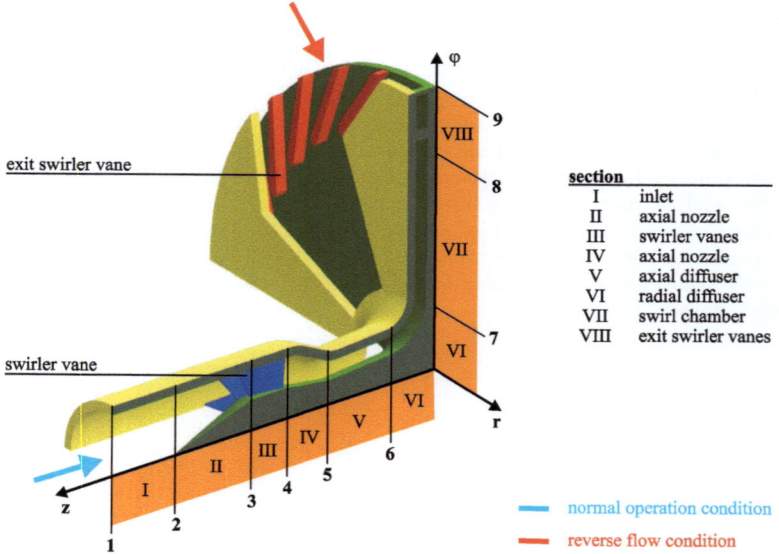

Figure 12.3: Quarter section of the backflow limiter with the several sections and positions for the 1D analysis.

The backflow limiter is divided into the following flow sections: the inlet section (I), inlet nozzle (II), inlet swirler vanes (III), second axial nozzle (IV), axial diffuser (V), radial diffuser (VI), swirl chamber (VII) and exit swirler vanes (VIII). To determine the pressure loss for each section, the static pressures and the dynamic pressures are calculated for each position between the different sectors marked in Figure 12.3 with position numbers 1 to 9.

For regular operation condition (blue arrow in Figure 12.3) the flow is accelerated in the inlet nozzle before the inlet swirler vanes impose a small swirl on the flow. Behind the vanes, the flow is accelerated again in a second nozzle before it enters the axial diffuser and is slowed down in the radial diffuser. The inlet swirl shall avoid flow separation in the radial diffuser or in the swirl chamber in regular flow direction. In reverse direction (red arrow), a swirl is induced by the exit swirler vanes (marked red in the cut-out of the swirl chamber in Figure 12.3). The swirl reaches a much higher intensity due to the contraction of the swirl chamber. Centrifugal forces push the flow radially outward and thus result in a much smaller effective cross section at position number 5.

Position number	1	2	3	4	5	6
D_i [m]	0	0	0.1	0.1	0.08	0.09
D_o [m]	0.2	0.2	0.2	0.2	0.14	0.16
z-coordinate [m]	0.7	0.59	0.45	0.38	0.3	0.2
Position number	7	8	9			
r-coordinate [m]	0.14	0.37	0.48			
Height [m]	0.02	0.02	0.02			

Table 12.2: Dimensions of the backflow limiter Mk. 4 design for a cylindrical coordinate system.

This results in a high pressure loss, i.e. small flow rates at a given pressure drop. Moreover the destruction of the swirl at the trailing edges of the inlet swirler vanes gives a very high additional flow resistance. The dimensions of the final design of the backflow limiter (Mk. 4 design) are listed in Table 12.2 in meter. For each position in the axial direction (position number 1 to 6) the inner diameter (D_i) and the outer diameter (D_o) are given with respect to the position in axial direction (z-coordinate). For each position in the radial direction (position number 7 to 9), the corresponding radius (r-coordinate) and height are given. The backflow limiter version shown in Figure 12.3 is designed using these data.

For the Mk. 4 design, the pressure loss coefficients ζ_i for the different sections i are determined using Equation 3.3 on page 28. A summary of the coefficients can be found in Table 12.3.

Section number	I	II	III	IV	V	VI	VII	VIII
Loss coefficient ζ	0.031	0.042	0.303	0.013	0.018	0.116	0.179	0.822

Table 12.3: Pressure loss coefficient ζ_i for the different sections number I to VIII for the Mk. 4 design with a preswirler angle of 10°.

12.2 Optimization Procedure

The HPLWR design concept assumes a coolant mass flow of 1160 kgs^{-1} (Squarer et al. [83]) at a pressure of 25 MPa and an inlet temperature of 280 °C. The feedwater flow is evenly divided between the four inlets of the RPV giving a mass flow rate of 290 kgs^{-1} per inlet line. The inlet pipe which precedes the backflow limiter has a diameter of 0.2 m resulting in an inlet velocity of 11.88 ms^{-1}.

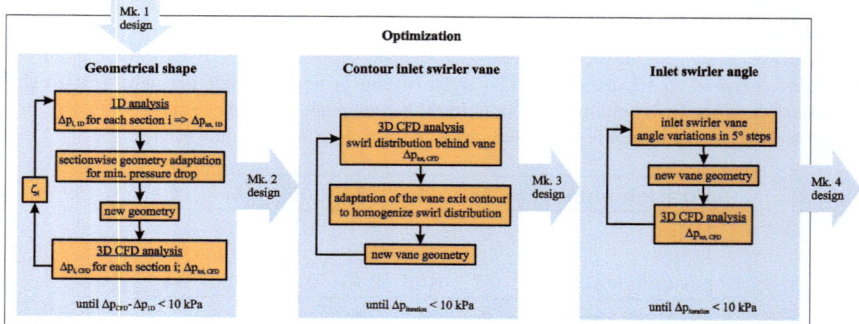

Figure 12.4: Scheme of the optimization process for the backflow limiter applying combined 1D and 3D CFD analyses.

Several analyses have been performed to obtain the described design, the different steps are summarized in the optimization scheme of Figure 12.4. In a preliminary analysis, the vane angles for inlet and exit swirlers are determined using a 1D analysis as a first design step. The steady state 1D momentum equation for incompressible fluids in accordance with the conservation of angular momentum as described in Oertel and Boehle [62] is used to determine the velocities and pressures for the different positions 1 to 9. The pressure drop for each section is estimated by applying general loss coefficients ζ used for turbo engine design (Sigloch [81]) as described in equation 3.3, on page 28.

The vane angles for inlet and exit swirlers are fixed for the following optimization of the geometry of the backflow limiter, especially of the nozzles and diffusers (shown on the left side ot the scheme in Figure 12.4). The resulting design from the preliminary analysis is called Mk. 1. For the first optimization step, geometric parameters of each section are changed to minimize the losses in regular operation condition. Then a 3D CFD analysis is performed with the new geometry. The obtained pressure losses are transferred back into new starting pressure loss coefficients ζ for another 1D flow optimization. This iterative process is executed until the difference between two steps is less then 10 kPa. The resulting optimized design is referred to as Mk. 2.

It is important for a low resistance in regular operation condition that the swirl created by the inlet swirler vanes is uniform over the cross section of the channel through the backflow limiter. The swirl S is calculated using equation 3.4, on page 28. The exit angles of the swirler vanes along the radial axis are modified to obtain a uniform profile. With the optimized swirl profile of this design named Mk. 3, the swirler vane angles are then tuned to obtain the target swirl and thus to minimize the losses in regular operation condition. Several geometries with varying angle positions are compared using the 3D CFD analysis resulting in the final design Mk. 4 which is shown in Figure 12.3.

12.3 Numerical Model in STAR-CD

The commercial CFD-code STAR-CD 3.26 has been used to perform the steady state analyses, where the discretized equations are solved adopting the SIMPLE algorithm.

12.3 Numerical Model in STAR-CD

The algebraic multigrid (AMG) method is used as a preconditioner for the CG-Solver to ease the convergence for the pressure correction equation. As spatial discretization method for the flow variables the QUICK scheme is applied.

The imported CATIA surface geometry of the backflow limiter can be seen in Figure 12.5. To ease the computational effort only a segment of the 3D geometry from the CAD-design is imported and used as the computational domain. For the axi-symmetric component it is sufficient to model a segment of 36° including one pre-swirler vane and three swirler vanes as shown in Figure 12.5. In this case appropriate boundary conditions have to be considered for the segment cutting planes. The pro-am generated mesh consists of hexahedral and trimmed cells in the fluid domain and is surrounded by several layers of refined prism cells to adjust the dimensionless wall distance y^+ for the application of high-Reynolds-turbulence models in combination with laws-of-the-wall. A grid of 1103393 cells in total with 541435 trimmed cells and 561958 prism cells has been found to meet the appropriate y^+ values (see section 5.7 on page 52).

The k-ω SST turbulence model in its high Reynolds formulation together with the standard-wall function is chosen according to the evaluation of turbulence models in section 5.8.

Boundary Conditions

The definition of the initial conditions for the flow field of the backflow limiter follows a constant distribution of all flow variables throughout the flow domain, since there is no information available for the real flow distribution at the beginning. In this case the stagnant conditions of a pressure vessel at 25 MPa are used for the initial flow field. The velocity throughout the domain is therefore zero in every cell volume (zero-flow condition).

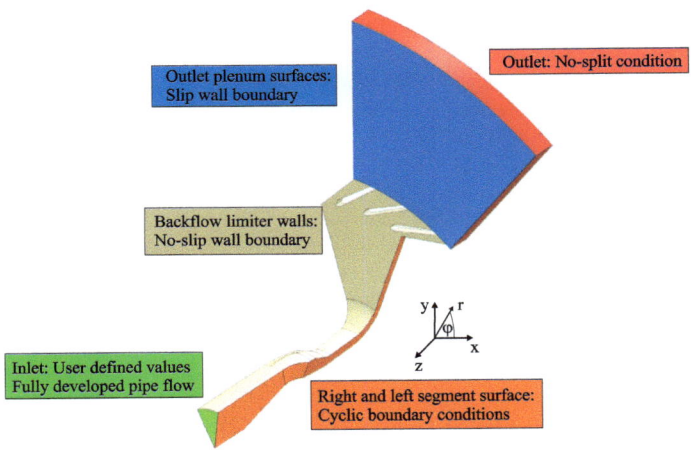

Figure 12.5: Numerical model of the backflow limiter with applied boundary conditions for the normal operation condition.

The following boundary conditions are implemented for the numerical domain of the backflow limiter and are illustrated in Figure 12.5 for the normal operation condition. The inlet mass flow with the velocity in axial direction (z-direction) is known, therefore a prescribed inlet flow is chosen using the Dirichlet boundary condition. Additionally, the turbulence properties for the kinetic energy k and the turbulent frequency ω have to be defined at the inlet. The distribution of those properties has to be consistent with the velocity profile. For this purpose a fully developed pipe flow is pre-calculated using the k-ω SST turbulence model and the prescribed mass flow and mapped on the inlet boundary. A block-structured mesh has been chosen for the calculation of the pipe flow. The values for the velocity in axial direction and the turbulent quantities are extracted from the separate calculation and mapped on a structured cell layer with the same topology, which is coupled to the inlet region of the backflow limiter. To avoid interference of the inlet nozzle section with the inlet domain an additional inlet section is added between the nozzle and the inlet boundary layer.

In the case of the backflow limiter a swirling flow exists behind the exit of the swirler vanes, where the pressure distribution on the outlet boundary is strongly influenced by the swirl and cannot be defined independently. To avoid this effect, the position of the outlet boundary is relocated using an outlet plenum. Its height is three times the height of the swirl chamber outlet; the shape features a segment angle of 36°. The outer diameter has been defined to fulfill the outlet condition that the gradients of the flow are close to zero at the boundary. The plenum consists of a coordinate-based block-structured hexahedral mesh, which is coupled to the pro-am mesh at the several outlets. A mesh with radially outward stretching cells is chosen to artificially increase the numerical diffusion and enforce the reduction of the flow gradients. For the single outlet plane of the plenum the no-split condition as implemented in STAR-CD is chosen.

The two cutting surfaces of the segment are defined as a cyclic pair, meaning that the flow repeats itself at the boundary. The mapping of the different variables from one surface to the other is done in an arbitrary way, since the computational grid of the backflow limiter is unstructured and the grid points from one surface do not exactly match those of the other one.

For the velocity components u_i at the walls of the flow limiter, the no slip condition ($u_i = u_{i,\text{wall}}$) is applied as the wall boundary condition with the default setting for roughness of the flow in pipes used in STAR-CD. The remaining outlet plenum surfaces are defined as slip wall boundaries to reduce interference of the walls with the outlet flow of the backflow limiter.

For the analysis in reverse flow direction, the inlet and outlet boundaries are switched and the former outlet plenum is reduced in diameter to minimize the cell number of the mesh, but is kept big enough to avoid flow interference of the swirler inlets with the inlet domain. In this case the prescribed flow option is used for the inlet boundary. The grid is adapted to meet the appropriate y^+ values for the occurring higher velocities which results in a higher total number of 1408345 cells.

12.4 Results of the Optimization

The post-processing method described in section 5.7, page 52 is applied to determine the individual pressure losses for the sections I to VIII. The same method is used to determine

12.4 Results of the Optimization

the mean swirl at the evaluation positions 1 to 9 (see Figure 12.3 on page 106).

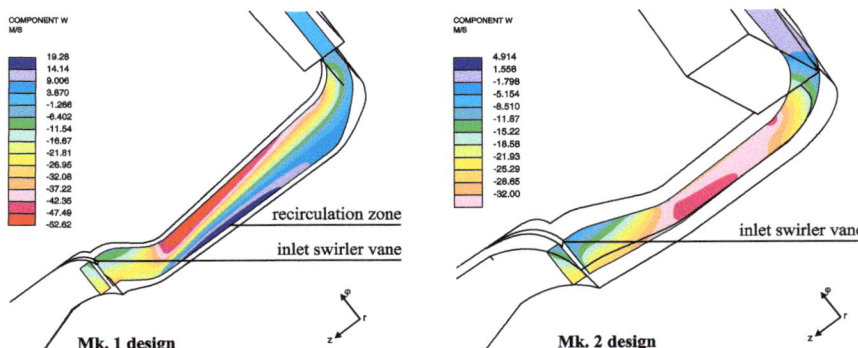

Figure 12.6: Velocity distribution in z-direction for one cutting plane of the Mk. 1 design (left side) and the Mk. 2 design (right side).

For the first step of the optimization process, the improved geometry of the backflow limiter (Mk. 1) from the 1D calculation is analyzed using CFD. Figure 12.6 shows the velocity profile in z-direction for one cutting plane in the middle of the segment. It can be seen that a recirculation zone is formed in the vicinity of the shaft throughout section V, starting at position 5. This causes the fluid to flow through a very narrow ring-shaped area at the outer diameter of the diffuser creating a high overall pressure drop of 1.55 MPa. The shape of section IV, V and VI is changed in several iterative steps to yield a minimum total pressure loss.

This trend is also illustrated in Figure 12.7 where the related pressure drop $\Delta p/p_{tot}$ is compared for positions number 1 to 9 for the design Mk. 1 and the optimized design Mk. 2. A peak for the related pressure loss can be observed between position 3 and 5 for the Mk. 1 design which can be attributed to the described recirculation zone. By changing the shape of the axial diffuser in section V and shortening its flow length the recirculation zone is omitted and the observed peak is reduced significantly. Additional optimizations have been performed for both axial nozzles and the radial diffuser. For the latter, the optimum shape for a minimum related pressure loss is a constant area bend which on the other side has a negative influence on the created swirl and therefore on the related pressure loss of the exit swirler section. The shape of the radial diffuser is balanced against an increasing related pressure loss for the exit swirler in section VIII to reduce the overall pressure loss. In Figure 12.6 on the right side the optimized design Mk. 2 is shown, which prevents the creation of a recirculating flow and features a reduced overall pressure loss of 0.344 MPa.

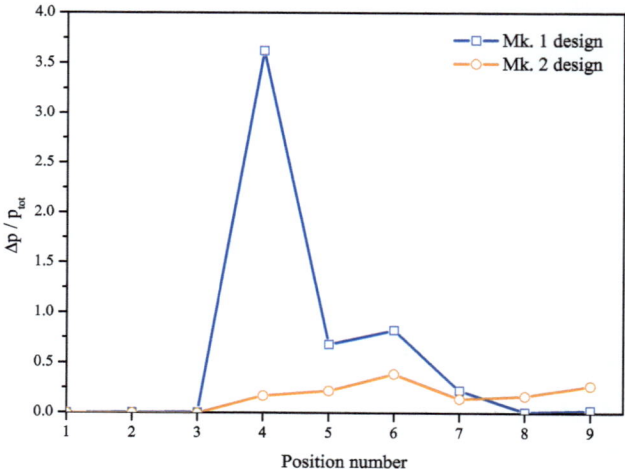

Figure 12.7: Comparison of the related pressure drop (ordinate) for positions number 1 to 9 (abscissa) for the Mk. 1 and Mk. 2 design.

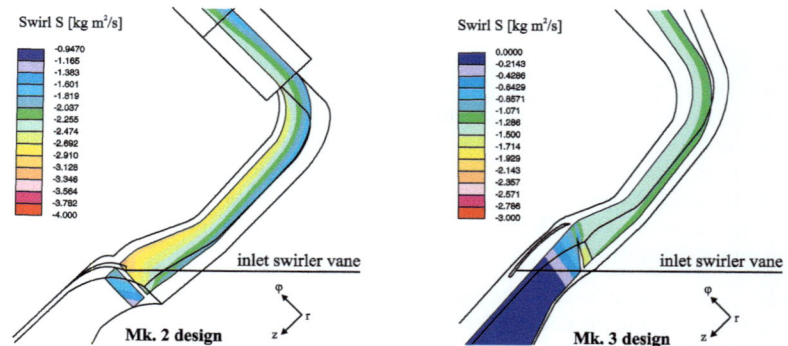

Figure 12.8: Calculated swirl distribution behind the inlet swirler vanes with an angle of 25° for the Mk. 2 (left side) and Mk. 3 design (right side).

In the next step, the contour of the inlet swirler vane is optimized in relation to the created swirl. In Figure 12.8 on the left side the calculated swirl distribution in z-direction behind one swirler vane for the Mk. 2 design is shown. The vane features an exit angle of 25° over the complete height. This design results in a non-uniform distribution of the swirl behind the trailing edges of the vanes in radial direction. To improve the performance, a vane with varying angle over the height is chosen, so that the condition for a potential flow vortex is fulfilled. Figure 12.8 shows on the right side the Mk. 3 design featuring such a vane contour. The swirl distribution in this case is very uniform and has the same

12.4 Results of the Optimization

value along the trailing edge of the vane. The Mk. 3 design results in a lower swirl along the flow path of the backflow limiter, therefore reducing the overall pressure loss to a value of 0.186 MPa (Figure 12.9).

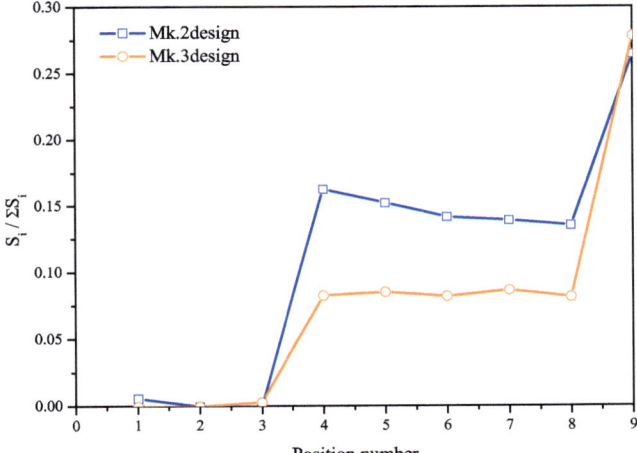

Figure 12.9: Comparison of the related swirl (ordinate) for positions number 1 to 9 (abscissa) for the Mk. 2 and Mk. 3 design.

A parametric study is performed to determine the best angle of the inlet swirler which gives the lowest pressure drop for the envisaged exit swirl of 60°. Several angles in steps of 5° are compared to find the design with the lowest pressure drop. Figure 12.10 gives the related pressure drop for the various angles for each position 1 to 9. Note that the pressure losses between positions 3 to 7 decrease with smaller angles, while the pressure loss at the entrance to the exit swirler vanes (position 8) increases. Quantitatively, the pressure losses between Positions 3 to 7 are higher and therefore dominant for the pressure loss improvement. A minimum overall pressure loss of 0.13 MPa can be found for a vane angle of 10°. This design is referred to as Mk. 4. For values below that margin, the swirl is too weak to have a positive effect on the flow and the overall losses increase again. Moreover, flow separation in the radial diffuser has to be expected then. Figure 12.11 illustrates the velocity profile for the r-ϕ plane at the middle of the swirl chamber for an inlet swirler vane angle of 10°. It can be noted, that the flow is smoothly redirected into the inlets between the different swirl vanes.

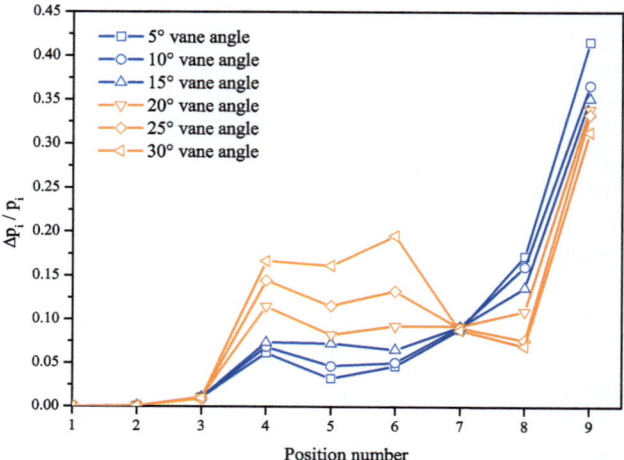

Figure 12.10: Related pressure drop for different swirler angles (ordinate) for position number 1 to 9 (abscissa).

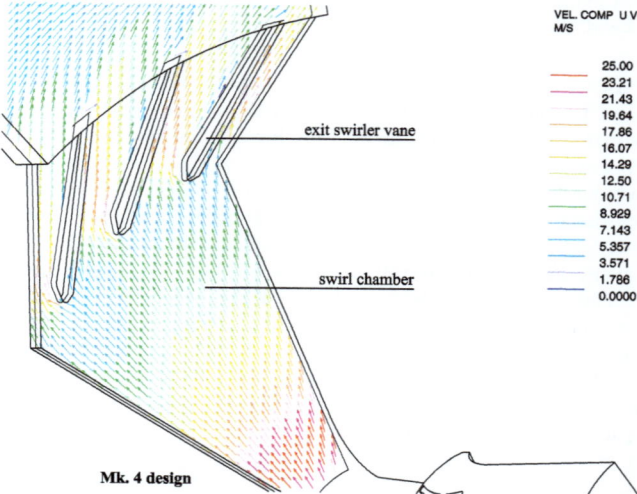

Figure 12.11: Vector plot of the velocity in the r-ϕ plane at the middle of the swirl chamber for the Mk. 4 design.

12.5 Characteristic of the Backflow Limiter

The Mk. 4 design is used for the 3D CFD analysis in reverse flow direction. The prescribed flow at the inlet, which is defined as described in section 12.3 is set to 290 kgs^{-1}. The resulting flow pattern for the velocity in z-direction can be seen in Figure 12.12. Three recirculation zones can be observed, which are marked as recirculation

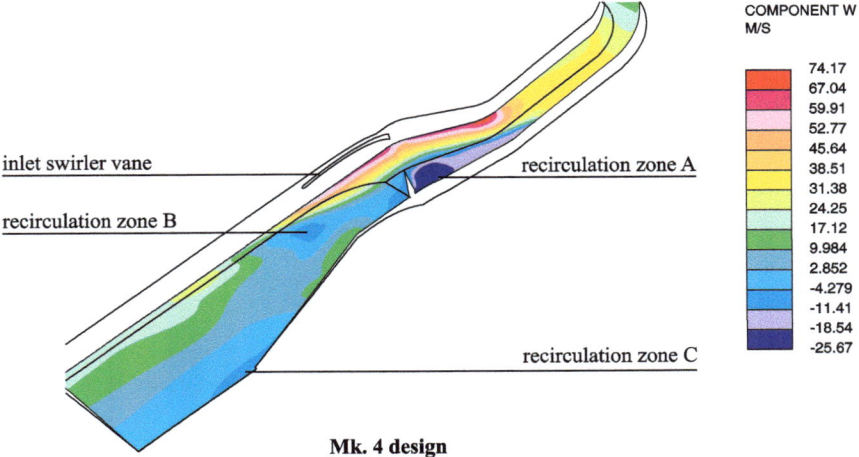

Figure 12.12: Velocity distribution in z-direction for one cutting plane for the Mk. 4 design in reverse flow direction.

zones A, B, and C, respectively. Zone A causes a large reduction of the effective flow cross section in the vicinity of position 5, so that the main flow is only allowed to pass at the outer diameter causing a high pressure loss, in this case 1.53 MPa. The second major pressure loss of 1.43 MPa occurs between the inlet swirler vanes caused by recirculation zone B. The third zone at the end of the diffuser section (inlet nozzle section I for regular flow) gives only a minor additional resistance. Setting the overall pressure loss of 3.16 MPa against the pressure loss in regular operation condition by using the resistance coefficients K_A and K_B from Equation 3.1 (page 28), results in a performance factor Σ of 21. The mass flow reduction for a comparable pressure drop in both direction gives a value of 5 or 80 %.

The maximum mass flow for the reverse direction is restricted by the critical mass flow in the smallest cross section. For supercritical water, evaporation and a two-phase flow will occur if the pressure is lower than 6.4 MPa at a temperature of 280 °C (Wagner and Kruse [94]), resulting in choking of the flow. Therefore, the reverse flow is limited by the critical mass flow as the largest possible mass flow for a pressure drop over 18.6 MPa. For the following parametric study, this maximum pressure drop gives the limit for the maximum mass flow.

Using the Mk. 4 design, different mass flows are calculated for both, regular operation condition and reverse flow direction. Figure 12.13 shows the corresponding pressure losses by its square root for different mass flows in both directions. A linear behavior can be

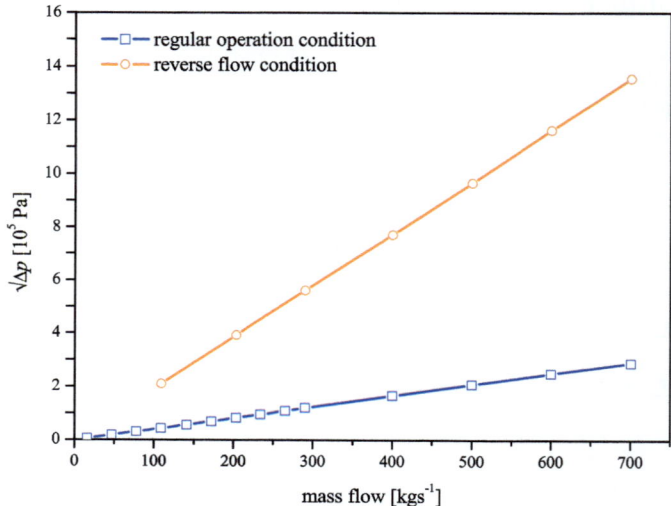

Figure 12.13: Square root of the total pressure loss versus the different mass flows in both directions for the Mk. 4 design.

observed over the whole investigated range. For a mass flow of 700 kgs^{-1}, a pressure drop of 18.45 MPa is observed. From these values, the resistance coefficients for both directions can be extrapolated using Equation 3.2. The Reynolds number as defined in equation 2.7 is used for the dimensionless representation of the characteristic of both coefficients K. Figure 12.14 shows the resistance coefficient K_B for various Reynolds numbers. The trend with a decreasing resistance coefficient for higher Reynolds numbers in an exponential manner can also be observed for applications using the simple vortex diode (Mueller [61]). The exponential fit for the available data points of the resistance coefficient K_B is given in Equation 12.1.

$$K_B(Re) = 2.6284 + 0.5364 \cdot e^{-(4.37 \cdot 10^{-7} \cdot Re)} \qquad (12.1)$$

Even the quite linear behavior for Reynolds numbers higher than 15×10^6 can be found in some cases. The lowest value for the resistance coefficient occurs for a Reynolds number of 17×10^6 in the vicinity of the normal inlet operation mass flow. For the reverse flow direction a clear trend compared to the simple vortex is not visible, as can be seen in Figure 12.15, where the resistance coefficient K_A is plotted versus different Reynolds numbers. This can be explained by the different working principles of the diodes, the backflow limiter develops its resistance mainly due to recirculation zones, while the simple diode uses swirl induced resistance. An average resistance value of 57.7 can be deducted from the available data points.

To evaluate the performance of the backflow limiter, the performance factor Σ is calculated applying the resistance coefficients K for corresponding Reynolds numbers (see Equation 3.1). In Figure 12.16 the performance factor is plotted versus different Reynolds

12.5 Characteristic of the Backflow Limiter

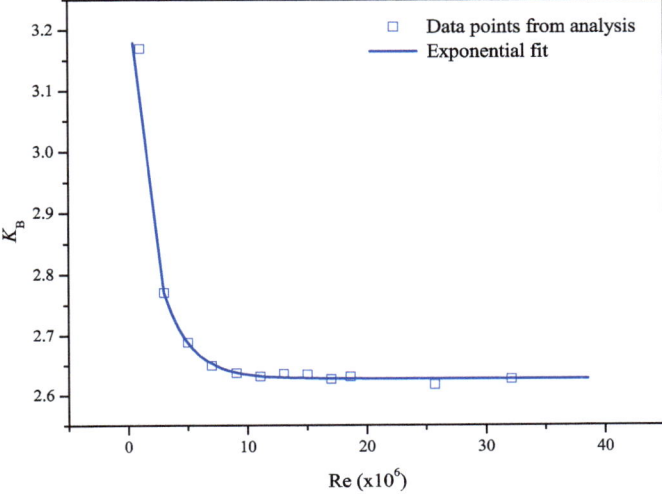

Figure 12.14: Resistance coefficient K_B versus different Reynolds numbers for the Mk. 4 design with applied exponential fit.

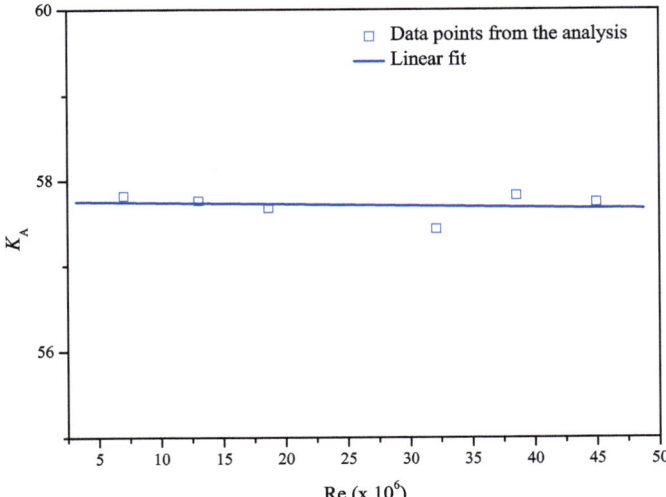

Figure 12.15: Resistance coefficient K_A versus different Reynolds numbers for the Mk. 4 design with applied linear fit.

numbers. A quite linear behavior can be observed, where the factor decreases with lower Reynolds numbers. Therefore, a linear fit has been applied to the available data points,

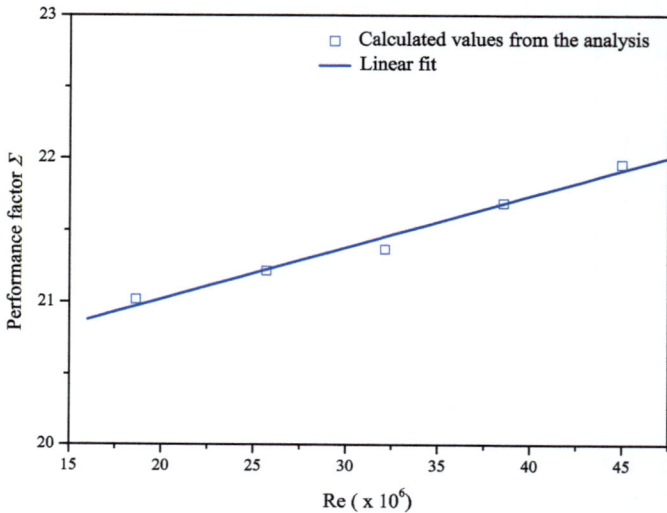

Figure 12.16: Performance factor Σ versus different Reynolds numbers and applied linear fit.

resulting in the following Equation:

$$\Sigma(Re) = 4 \cdot 10^{-13} \cdot Re + 20.295 \qquad (12.2)$$

An average value in the vicinity of 21.5 can be deducted for the investigated region.

In summary, a method to optimize the backflow limiter using a combination of 1D analyses and 3D CFD analyses is presented. Due to the iterative process it was possible to reduce the resistance in regular flow condition starting with the optimized 1D geometry by 90 %. For this purpose, the overall geometrical shape, the contour of the inlet swirler vanes, and their angle are adapted to reduce the resistance in regular operation condition.

The final design of the backflow limiter features 10 inlet swirler vanes with an angle of 10° and 30 exit swirler vanes with an angle of 60°; the swirl chamber has an overall diameter of 0.95 m. In case of a loss of coolant accident (LOCA), e.g. in case of a postulated break of one of the four inlet feedwater lines, the backflow limiter is able to reduce the mass flow in reverse direction by a factor of 5 or 80 %. A similar design, tested by Siemens KWU in Karlstein in Germany ([61]), featured a reduction of 90 %, but for lower pressure differences. The overall performance Σ of the backflow limiter varies between 21 and 22 for the investigated range of Reynolds numbers between 1×10^6 and 45×10^6.

13 Conclusions

The presented design of the RPV and internals for a supercritical water-cooled reactor demonstrates for the first time that such a reactor with the high operating pressure of 25 MPa and outlet temperatures around 500 °C is technically feasible. Several new design features have been developed to realize this task. An evolutionary approach featuring two and three heat up steps has been chosen to keep the cladding temperatures in the core inside the material limit. Starting with a one pass core, the design study is extended to a two pass and even a three pass core arrangement.

For this purpose a leak tight fuel assembly cluster has been designed for both, the two pass core with its co-current and counter-current flow arrangement and the three pass core with its evaporator and two superheater sections. Additional mixing plena above and below the core establish the necessary connection between the heat up stages and provide intermediate mixing of the coolant. The design assures that the total coolant flow path remains closed against leakage of colder moderator water even in case of large thermal expansions of the components. Several openings in the foot piece allow to adjust the moderator mass flow through the assembly to optimize the power distribution. Assembly clusters can be exchanged freely between the different regions of the core for the three pass arrangement facilitating the optimization of the power profile and burn up. For the exchange of fuel assembly clusters from the counter-current to the co-current region of the two pass core, the bushing of the head piece is exchanged and an additional cylindrical can is applied to the window element.

Due to the high operating pressure and temperature, special care has been contributed to the design of the internals, including the core barrel, the core support plate, and the control rod drive and its guide tubes. They are using state of the art technologies of pressurized water reactors and have been dimensioned accordingly for the challenging conditions using the safety standards of the nuclear safety standards commission (KTA) in Germany. Modifications include the implementation of the different flow paths for the two and three pass core arrangement and adaptation to the occurring higher thermal deformations. As an alternative to a conventional PWR design with reflector water and thermal shield, a steel reflector is applied which gives a flattened power profile and reduces neutron leakages from the core.

To account for the higher steam outlet temperatures and the individual flow path in the core, the steam plenum, the outlet tubes and the lower mixing chamber are designed differently from known LWR designs. The steam plenum features extractable hot tubes and is fixed with support brackets at the circumference of the inner RPV wall though reducing thermal stresses and deformations between the components. The internals, together with the extractable hot pipes allow for easy access to the core and the fuel assembly clusters, reducing the complexity of loading and unloading during revisions. The design with a shortened steam plenum provides space for an in-vessel accumulator for coolant above the core giving additional safety in a loss of coolant accident.

To ensure the required leak tightness, replaceable c-ring seals have been applied for the connection between the steam plenum, the hot tubes, and the penetrating head pieces of the fuel assembly clusters. To verify the sealings, all fuel assembly openings and exit nozzles have been checked for allowable deformations using finite element methods. The coupled thermal-mechanical analysis allowed to resolve the local deformations for each fuel assembly opening, yielding maximum deviations of 0.275 mm from a preferable ring shape. Those deformations stay well within the tolerance limit of the applied c-ring sealings with 0.3 mm. To reduce occurring deformations for the exit nozzles a decoupling from the guide strips and a conical stiffening is chosen. The occurring deformations for the exit nozzles have been found to be negligible therefore ensuring leak tightness of the applied C-ring sealings. The analysis also allowed to determine occurring mechanical and thermal stresses for the whole casing, yielding acceptable stress intensities throughout the component.

Those internals have been incorporated within the scope of this study into a design for the reactor pressure vessel with closure head for all three core designs. The design for the one and two pass core has been optimized and dimensioned to contain a core with 88 assembly clusters using the mentioned safety standards. The inner diameter of the vessel for both core designs is 3.38 m, the total height about 13 m; the maximum thickness of the shell is 0.51 m. The dimensioning has been verified using finite elements analyses which included mechanical and thermal loads. As a consequence of the three pass design, the diameter of the core and therefore the diameter of the reactor pressure vessel were increased. To minimize the wall thickness in the vicinity of the flanges, a mechanical finite element analysis has been performed to receive acceptable stress intensities in those regions. With an inner vessel diameter of 4.47 m and a maximum wall thickness of 0.45 m in the cylindrical part, the pressure vessel for the three pass core can still be forged and allows to use known manufacturing technology.

Existing vessel materials have been used for all designs due to the implementation of a thermal sleeve at the outlet. The performed coupled thermal-mechanical analysis using finite elements states that the calculated mass flow for the thermal sleeve is sufficient to isolate the high temperatures of the outlet pipes from the outlet flanges of the reactor pressure vessel for all three designs. Additionally, the thermal deformations due to the different temperatures for the inlet and outlet flange are small enough to ensure leak tightness of the two o-ring seals for the closure head.

The RPV design has also been evaluated for necessary safety components in the case of transients. One of the most important tasks is the permanent cooling of the core for all conditions. A backflow limiter, which is also considered for advanced boiling water reactors, has been chosen to minimize feedwater losses in the case of a break of the feedwater line. For this passive component, activation by instrumentation and control systems is not required and operation is independent of power supply. Starting from a simple vortex diode, a method to optimize this backflow limiter for the implementation in the feedwater line, using a combination of 1D analyses and 3D CFD analyses, is presented. Starting with the optimized 1D geometry it was possible to reduce the resistance in regular flow condition by 90 % using this iterative method. For this purpose, the overall geometrical shape, the contour of the inlet swirler vanes, and their angle were adapted to reduce the resistance in regular operation condition.

The final design of the backflow limiter features 10 inlet swirler vanes with an angle of 10°, and 30 exit swirler vanes with an angle of 60°; the swirl chamber has an overall diameter of 0.95 m. The mass flow in the event of a postulated break of one of the four feedwater lines is reduced to approximately 20 % of the value, which would result without the backflow limiter. CFD analyses have been used to predict the flow characteristic of the backflow limiter for both directions and a certain range of Reynolds numbers applying the resistance coefficient ζ. The resulting performance can also be expressed by the performance factor Σ which in this case yields values between 21 and 22. Comparable mass flow reduction and performance can be found for flow restrictors, which are implemented in the advanced boiling water reactor SWR 1000 (Pasler [68], Mueller[61]).

Nevertheless, several design details still need to be added. An essential requirement for the feasibility of the three pass core design is an excellent coolant mixing in the mixing chambers above and below the core. Efficient mixing devices for the upper mixing chamber have to be identified and worked out in detail to decide about the optimum configuration. The same approach has to be chosen for the presented swirlers in the lower mixing plenum. For the core itself, a thermal insulation of the assembly and water boxes will be advisable to minimize heat up of the moderator water; orifices need to be added at the evaporator inlet to avoid density wave oscillations.

The investigation of the RPV design has to be extended to load classes which include transients, like changes to the system for functionally fit conditions as start-up, full-load operation and shutdown. The required heat transfer coefficients have to be determined using detailed flow analyses to maintain the temperature characteristic for different conditions.

The optimization of the backflow limiter design does not yet include mechanical analyses. The impact of the occurring pressure forces on the structural design have to be determined and evaluated. Possible design changes will also influence the behavior and performance of the component.

After being reviewed from all partners of the project in September 2007, the design of the RPV and internals is now available for detailed analyses of the core and the reactor by the different working package partners. The presented work represents an important contribution for the final assessment of this new reactor type. Additionally, the flow characteristic of the backflow limiter is provided as an input for safety system analyses.

Bibliography

[1] N. AKSAN, T. SCHULENBERG, D. SQUARER, X. CHENG, D. STRUWE, V. SANCHEZ, P. DUMAZ, R. KYRKI-RAJAMAKI, D. BITTERMANN, A. SOUYRI, Y. OKA, AND S. KOSHIZUKA, *Potential Safety Features and Safety Analysis Aspects for High Performance Light Water Reactor (HPLWR)*, no. 1223, GENES4/ANP2003, Kyoto, Japan, 2003.

[2] T. R. ALLEN, L. TAN, Y. CHEN, X. REN, K. SRIDHARAN, G. S. WAS, G. GUPTA, AND P. AMPORNRAT, *Corrosion of Ferritic-Martensitic Alloys in Supercritical Water for GenIV Application*, no. 419, Proceedings of GLOBAL 05, Tsukuba, Japan, 2005.

[3] T. R. ALLEN, L. TAN, Y. CHEN, K. SRIDHARAN, M. T. MACHUT, J. GAN, G. GUPTA, G. S. WAS, AND S. UKAI, *Corrosion and Radiation Response of Advanced Ferritic-Martensititc Steels for Generation IV Application*, no. IL001, Proceedings of GLOBAL 05, Tsukuba, Japan, 2005.

[4] O. ANTONI AND P. DUMAZ, *Preliminary Calculations of a Supercritical Light Water Reactor Concept using the CATHARE Code*, no. 3146, Proceedings of ICAPP'03, Cordoba, Spain, 2003.

[5] P. BAKER, *A Comparison of Fluid Diodes*, no. D6, Second Cranfield Fluidics Conference, Cambridge, United Kingdom, 3rd-5th January 1967.

[6] K.-J. BATHE, ed., *Finite-Elemente-Methoden*, Springer Verlag, Berlin, Germany, 2002.

[7] D. BITTERMANN, D. SQUARER, T. SCHULENBERG, AND Y. OKA, *Economic Prospects of the HPLWR*, no. 1003, GENES4/ANP2003, Kyoto, Japan, 2003.

[8] D. BITTERMANN, D. SQUARER, T. SCHULENBERG, Y. OKA, P. DUMAZ, R. KYRI-RAJAMÄKI, N. AKSAN, C. MARACZY, AND A. SOUYRI, *Potential Plant Characteristics of a High Performance Light Water Reactor (HPLWR)*, no. 3137, Proceedings of ICAPP'03, Cordoba, Spain, 2003.

[9] D. BITTERMANN, J. STARFLINGER, AND T. SCHULENBERG, *Turbine Technologies for High Performance Light Water Reactors*, no. 4195, Proceedings of ICAPP'04, Pittsburg, United States, 2004.

[10] W. BRETTSCHUH, *Experimental Verification of SWR 1000 Passive Components and Systems*, no. 313, Proceedings of the Annual Meeting on Nuclear Technology, Karlsruhe, Germany, 2007.

[11] J. BUONGIORNO, *The Supercritical Water Cooled Reactor: Ongoing Research and Development in the U.S.*, no. 4229, Proceedings of ICAPP'04, Pittsburgh, United States, 2004.

[12] J. BUONGIORNO AND P. MACDONALD, *Supercritical Water Reactor (SCWR), Progress Report for the FY-03 Generation-IV R & D Activities for the Development of the SCWR in the U.S.*, Tech. Report INEEL/EXT-03-01210, Idaho National Engineering and Environmental Laboratory (INEEL), 2003.

[13] S. BUSHBY, *Conceptual Designs for Advanced, High-Temperature CANDU Reactors*, no. ICONE-8470, Proceedings of ICONE-8, ASME, 2000, pp. 1–7.

[14] M. CASEY, *Thematic Area 6: Best Practice Advice for Turbomachinery Internal Flows*, QNET-CFD Network Newsletter, 2 (2004), pp. 40–46.

[15] M. CASEY AND T. WINTERGERSTE, *ERCOFTAC Best Practice Guidelines*, tech. report, European Reasearch Community on Flow, Turbulence and Combustion, 2000.

[16] M. CERVANTES AND F. ENGSTRÖM, *Eddy Viscosity Turbulence Models and Steady Draft Tube Simulations*, no. Paper 2, Proceedings of the third IAHR/ERCOFTAC workshop on draft tube flow, 2005, pp. 37–44.

[17] X. CHENG, T. SCHULENBERG, D. BITTERMANN, AND P. RAU, *Design Analysis of Core Assemblies for Supercritical Pressure Conditions*, Nuclear Engineering and Design, 223 (2003), pp. 279–294.

[18] S. CHURCHILL AND H. CHU, *Correlating Equations for Laminar and Turbulent Free Convection from a Vertical Plate*, International Journal for Heat and Mass Transfer, 18 (1975), pp. 1323–1329.

[19] R. COURANT, E. ISAACSON, AND M. REES, *On the Solution of Nonlinear Hyperbolic Differential Equations by Finite Differences*, Communications On Pure & Applied Mathematics, 5 (1952), pp. 243–255.

[20] K. DOBASHI, Y. OKA, AND S. KOSHIZUKA, *Conceptual Design of a High Temperature Power Reactor Cooled and Moderated by Supercritical Light Water*, Annual Nuclear Energy, 25 (1998), pp. 487–505.

[21] DUBBEL, ed., *Taschenbuch fuer den Maschinenbau*, W. Beitz and K.-H. Grote, 2001.

[22] K. EHRLICH, J. KONYS, AND L. HEIKINHEIMO, *Materials for High Performance Light Water Reactors*, no. 3310, Proceedings of ICAPP'03,Cordoba, Spain, 2003.

[23] J. FERZIGER AND M. PERIĆ, eds., *Computational Methods for Fluid Dynamics*, Springer Verlag, Berlin, 1996.

[24] K. FISCHER, E. GUELTON, AND T. SCHULENBERG, *Festigkeitsanalyse des Reaktordruckbehaelters fuer einen Leichtwasserreaktor mit ueberkritischen Dampfzustaenden*, no. 723, Proceedings of the Annual Meeting on Nuclear Technology, Karlsruhe, Germany, 2007.

[25] K. FISCHER, E. LAURIEN, A. CLASS, AND T. SCHULENBERG, *Design and Optimization of a Backflow Limiter for the High Performance Light Water Reactor*, no. 175767, Proceedings of GLOBAL 07, Boise, United States, 2007.

[26] K. FISCHER, E. LAURIEN, A. CLASS, AND T. SCHULENBERG, *Hydraulic Analysis of a Backflow Limiter for the High Performance Light Water Reactor*, to be published in Proceedings of the Annual Meeting on Nuclear Technology, Hamburg, Germany, 2008.

[27] K. FISCHER, T. REDON, G. MILLET, C. KOEHLY, AND T. SCHULENBERG, *Thermo-Mechanical Stress and Deformation Analysis of a HPLWR Pressure Vessel and Steam Plenum*, no. 8050, to be published in Proceedings of ICAPP'08, Anaheim, United States, 2008.

[28] K. FISCHER, T. SCHNEIDER, T. REDON, T. SCHULENBERG, AND J. STARFLINGER, *Mechanical Design of Core Components for a High Performance Light Water Reactor with a Three Pass Core*, no. 175772, Proceedings of GLOBAL 07, Boise, United States, 2007.

[29] K. FISCHER AND T. SCHULENBERG, *Auslegung eines Reaktordruckbehaelters mit Kerneinbauten fuer einen Leichtwasserreaktor mit ueberkritischen Dampfzustaenden*, Student Workshop of the Annual Meeting on Nuclear Technology, Aachen, Germany, 2006.

[30] K. FISCHER, J. STARFLINGER, AND T. SCHULENBERG, *Conceptual Design of a Reactor Pressure Vessel and its Internals for a HPLWR*, no. 6098, Proceedings of ICAPP'06, Reno, United States, 2006.

[31] FRAMATOME ANP, *Technology of the European Pressurized Water Reactor (EPR)*, tech. report, Framatome ANP, Paris, France, 2005.

[32] R. GALLAGHER, ed., *Finite-Element-Analysis*, Springer Verlag, Berlin, Germany, 1976.

[33] GENERAL ELECTRIC, *Supercritical Pressure Power Reactor, A Conceptual Design*, Tech. Report HW-59684, Atomic Energy Commission Research and Development report, Hanford Laboratories, General Electric, 1959.

[34] P. GEORGE, J. WARD, AND F. MITCHELL, *Vortex Diode Characteristics at High Pressure Ratios*, no. 10, Proceedings of the Instrumentation and Measurements Control Meeting, London, United Kingdom, 1975.

[35] GFD - GESELLSCHAFT FÜR DICHTUNGSTECHNIK MBH, *Metall - O - Ringe und C - Ringe für extreme Betriebsbedingungen*, tech. report, 2005.

[36] V. GNIELINSKI, *Ein neues Berechnungsverfahren für die Waermeuebertragung im Uebergangsbereich zwischen laminarer und turbulenter Rohrstroemung*, Forschung im Ingenieurwesen, 61 (1995), pp. 240–248.

[37] E. GUELTON AND K. FISCHER, *Festigkeitsanalyse fuer den Reaktordruckbehaelter des High Performance Light Water Reactor (HPLWR)*, Wissenschaftliche Berichte FZKA 7270, Forschungszentrum Karlsruhe GmbH, Karlsruhe, Germany, 2006.

[38] H. HOFFMANN, *Experimental Investigations into Coolant Cross Mixing and Pressure Drop in Multi-Rod Bundles with Helical Type Spacers*, Tech. Report KFK-1843, Gesellschaft fuer Kernforschung MbH, Karlsruhe, Germany, 1973.

[39] J. HOFMEISTER, E. LAURIEN, A. CLASS, AND T. SCHULENBERG, *Turbulent Mixing in the Foot Piece of a HPLWR Fuel Assembly*, no. 066, Proceedings of GLOBAL 05, Tsukuba, Japan, 2005.

[40] J. HOFMEISTER, T. SCHULENBERG, AND J. STARFLINGER, *Optimization of a Fuel Assembly for a HPLWR*, no. 5077, Proceedings of ICAPP'05, Seoul, Korea, 2005.

[41] J. HOFMEISTER, C. WAATA, J. STARFLINGER, , T. SCHULENBERG, AND E. LAURIEN, *Fuel Assembly Design Study for a Reactor with Supercritical Water*, Nuclear Engineering and Design, 237 (2007), pp. 1513–1521.

[42] W. HUFSCHMIDT AND E. BRUECK, *Der Einfluß temperaturabhaengiger Stoffwerte auf den Waermeuebergang bei turbulenter Stroemung von Fluessigkeiten in Rohren bei hohen Waermestromdichten und Pr-Zahlen*, International Journal for Heat and Mass Transfer, 11 (1968), pp. 1041–1048.

[43] T. ICHIMURA, S. UEDA, S. SAITO, AND T. OGINO, *Design Verification of the Advanced Accumulator for the APWR in Japan*, no. 8435, Proceedings of ICONE 8, Baltimore, United States, 2000.

[44] H. JOO, K. BAE, H. LEE, J. NOH, AND Y. BAE, *A Conceptual Core Design with a Rectangular Fuel Assembly for a Thermal SCWR System*, no. 5223, Proceedings of ICAPP'05, Seoul, Korea, 2005.

[45] K. KAMEI, A. YAMAJI, Y. ISHIWATARI, L. JIE, AND Y. OKA, *Fuel and Core Design of Super LWR with Stainless Steel Cladding*, no. 5527, Proceedings of ICAPP'05, Seoul, Korea, 2005.

[46] K. KATAOKA, S. SHIGA, K. MORIYA, Y. OKA, S. YOSHIDA, AND H. TAKAHASHI, *Progress of Development Project of Supercritical-water Cooled Power Reactor*, no. 3258, Proceedings of ICAPP'03, Cordoba, Spain, 2003.

[47] H. KHARTABIL, R. DUFFEY, N. SPINKS, AND W. DIAMOND, *The Pressure-Tube Concept of Generation IV Supercritical Water-Cooled Reactor (SCWR): Overview and Status*, no. 5564, Proceedings of ICAPP'05, Seoul, Korea, 2005.

[48] C. KING, *Power Fluidics for Nuclear and Process Plant Safety and Control*, no. C94/83, Proceedings Conference on Heat and Fluid Flow in Nuclear and Process Plant Safety, Institute of Mechanical Engineers, London, United Kingdom, 1983.

[49] P. KONAKOV, *Berichte der Akademie der Wissenschaften der UDSSR*, Band LI, 51 (1946), pp. 503–506.

[50] KTA-GESCHAEFTSSTELLE C/O BUNDESAMT FUER STRAHLENSCHUTZ (BFS), ed., *Reactor Pressure Vessel Internals*, Kerntechnischer Ausschuss (KTA), 1998, ch. 3204.

[51] KTA-GESCHAEFTSSTELLE C/O BUNDESAMT FUER STRAHLENSCHUTZ (BFS), ed., *Components of the Reactor Coolant Pressure Boundary of Light Water Reactors, Part 1: Materials*, Kerntechnischer Ausschuss (KTA), 2000, ch. 3201.1.

[52] KTA-GESCHAEFTSSTELLE C/O BUNDESAMT FUER STRAHLENSCHUTZ (BFS), ed., *Components of the Reactor Coolant Pressure Boundary of Light Water Reactors, Part 2: Design and Analysis*, Kerntechnischer Ausschuss (KTA), 2000, ch. 3201.2.

[53] B. LEONARD, *A Stable and Accurate Convective Modeling Procedure based on Quadratic Upstream Interpolation*, Computer Methods in Applied Mechanics and Engineering, 19 (1979), pp. 59–98.

[54] B. LEONHARD, *Order of Accuracy of QUICK and Related Convection-Diffusion Schemes*, Applied Mathematic Modeling, 19 (1995), pp. 640–653.

[55] P. MACDONALD (LEADING AUTHOR), *Feasibility Study of Supercritical Light Water Cooled Reactors for Electric Power Production*, Tech. Report DE-FG07-02SF22533, Nuclear Energy Research Initiative Project 2001-001, 2005.

[56] J. MARCHATERRE AND M. PETRICK, *Review of the Status of Supercritical Water Reactor Technology*, Tech. Report ANL-6202, Atomic Energy Commission Research and Development report, Argonne National Laboratory, 1960.

[57] F. MENTER, *Two-Equation Eddy-Viscosity Turbulence Models for Engineering Applications*, AIAA Journal, 32 (1994), pp. 1598–1605.

[58] G. MÜLLER AND C. GROTH, eds., *FEM fuer Praktiker*, expert Verlag, Renningen-Malmsheim, Germany, 1997.

[59] P. MOLLER, *A Radial Diffuser using Incompressible Flow between Narrowly Spaced Disks*, Journal of Basic Engineering, (1966), pp. 155–162.

[60] A. MOTAMED-AMINI AND I. OWEN, *The Expansion of Wet Steam Through a Compressible Confined Vortex in a Fluidic Vortex Diode*, International Journal of Multiphase Flow, 13 (1987), pp. 845–856.

[61] H. MUELLER, *Passive Fluidik-Elemente/Wirbelelemente: Experimentelle Bestimmung der Kennlinien und der Geschwindigkeitsverteilung fuer eine große Wirbeldrossel (Power Fluidic)*, Tech. Report E311/92/40, Siemens KWU, Erlangen, Germany, 1973.

[62] H. OERTEL JR. AND M. BOEHLE, eds., *Stroemungsmechanik*, F. Vieweg & Sohn Verlagsgesellschaft mbH, Braunschweig, 2002.

[63] H. OERTEL JR. AND E. LAURIEN, eds., *Numerische Stroemungsmechanik*, F. Vieweg & Sohn Verlagsgesellschaft mbH, Braunschweig, 2nd ed., 2003.

[64] Y. OKA, *Review of High Temperature Water and Steam Cooled Reactor Concepts*, no. 104, Proceedings of SCR2000 Symposium, The University of Tokyo, 2000.

[65] Y. OKA AND S. KOSHIZUKA, *Conceptual Design of a Supercritical-Pressure Direct-Cycle Light Water Reactor*, vol. 1, Proc. Int. Conf. Design and Safety of Advanced Nuclear Power Plants, Tokyo, Japan, 1992, pp. 1–7.

[66] Y. OKA AND S. KOSHIZUKA, *Concept and Design of a Supercritical-Pressure, Direct-Cycle Light Water Reactor*, Nuclear Technology, 103 (1993), pp. 295–302.

[67] I. OWEN AND A. MOTAMED-AMINI, *Compressible Flow Characteristics of a Vortex Diode Operating with Superheated Steam*, no. 86-WA/FE-5, Winter Annual Meeting of the American Society of Mechanical Engineers, Anaheim, United States, 1986.

[68] D. PASLER, *The Safety Concept of the SWR 1000*, no. 4207, Proceedings of ICAPP'04, Pittsburgh, United States, 2004.

[69] S. PATANKAR AND D. SPALDING, *A Calculation Procedure for Heat, Mass and Momentum Transfer in Three-Dimensional Parabolic Flows*, International Journal of Heat and Mass Transfer, 15 (1972), pp. 1787–1806.

[70] B. PETUKHOV AND L. ROIZEN, *Heat Transfer in a Single-Phase Medium under Supercritical Conditions*, High Temperature, 2 (1964), pp. 65–68.

[71] P. ROACHE, ed., *Verification and Validation in Computational Science and Engineering*, Hermosa Publishers, Albuquerque, United States, 1998.

[72] W. RODI, *Progress in Turbulence Modeling for Incompressible Flows*, no. Paper 81-45, AIAA, St. Louis, United States, 1981.

[73] W. SCHNELL, D. GROSS, AND W. HAUGER, eds., *Technische Mechanik, Band 2: Elastostatik*, Springer Verlag, Berlin, Germany, 1998.

[74] T. SCHULENBERG, K. FISCHER, AND J. STARFLINGER, *Review of Design Studies for High Performance Light Water Reactors*, no. SCR2007-P024, 3rd International Symposium on SCWR-Design and Technology, Shanghai, China, 2007.

[75] T. SCHULENBERG, J. HOFMEISTER, K. FISCHER, AND J. STARFLINGER, *Design Options for High Performance Light Water Reactors*, 28th Annual Conference of the Canadian Nuclear Society, Saint John, Canada, 2007.

[76] T. SCHULENBERG AND J. STARFLINGER, *Core Design Concepts for High Performance Light Water Reactors*, Nuclear Engineering and Technology, 39 (2007).

[77] T. SCHULENBERG AND J. STARFLINGER, *European Research Project on High Performance Light Water Reactors*, no. SCR2007-I001, 3rd International Symposium on SCWR-Design and Technology, Shanghai, China, 2007.

[78] T. SCHULENBERG, J. STARFLINGER, AND J. HEINECKE, *Three Pass Core Design Proposal for a High Performance Light Water Reactor*, 2nd COE-INES-2 International Conference on Innovative Nuclear Energy Systems, INES-2, Yokohama, Japan, 2006.

[79] B. SIDHU, N. SYRED, AND A. STYLES, *Flow and General Characteristics of High Performance Diodes*, Proceedings of the Winter Annual Meeting, Chicago, United States, 1980, pp. 121–131.

[80] E. SIEDER AND G. TATE, *Heat Transfer and Pressure Drop of Liquids in Tubes*, Industrial & Engineering Chemistry, 8 (1936), pp. 1429–1435.

[81] H. SIGLOCH, ed., *Stroemungsmaschinen - Grundlagen und Anwendungen*, Hanser Fachbuchverlag, München, Germany, 2004.

[82] V. SLIN, V. VOZNESSENSKSY, AND A. AFROV, *The Light Water Integral Reactor with Natural Circulation of the Coolant at Supercritical Pressure B-500 SKDI*, Nuclear Engineering and Design, 144 (1993), pp. 327–336.

[83] D. SQUARER, T. SCHULENBERG, D. STRUWE, Y. OKA, D. BITTERMANN, N. AKSAN, C. MARACZY, R. KYRKI-RAJAMÄKI, A. SOURYI, AND P. DUMAZ, *High Performance Light Water Reactor*, Nuclear Engineering and Design, 222 (2003), pp. 167–180.

[84] K. SRIDHARAN, A. ZILLMER, J. R. LICHT, T. R. ALLEN, M. H. ANDERSON, AND L. TAN, *Corrosion Behavior of Candidate Alloys for Supercritical Water Reactors*, no. 4136, Proceedings of ICAPP'04, Pittsburgh, United States, 2004.

[85] N. SYRED AND P. ROBERTS, *Use of Vortex Diodes Applied to Post Accident Heat Removal Systems*, no. 79-HT-9, ASME/AIchE 18th National Heat Transfer Conference, San Diego, United States, 1979.

[86] I. SZABÓ, ed., *Hoehere Technische Mechanik*, Springer Verlag, Berlin, Germany, 1977.

[87] S. TIMOSHENKO AND J. GOODIER, eds., *Theory of Elasticity*, McGraw-Hill, Singapore, 1982.

[88] S. TOWER, *1000MW$_e$ Supercritical Pressure Nuclear Reactor Plants Study*, Tech. Report WCAP-2042, Atomic Energy Commission Research and Development report, Westinghouse, 1962.

[89] K. TRAUBE AND L. SEYFFERTH, *Der Heissdampfreaktor - Konstruktion und Besonderheiten*, Atomwirtschaft, (1969), pp. 539–542.

[90] US DOE NUCLEAR ENERGY RESEARCH ADVISORY COMMITEE, *A Technology Roadmap for Generation IV Nuclear Energy Systems*, tech. report, Generation IV International Forum, 2002.

[91] VDI-GESELLSCHAFT VERFAHRENSTECHNIK UND CHEMIEINGENIEURWESEN, ed., *VDI-Waermeatlas*, Springer Verlag, Berlin, Germany, 8th ed., 1997.

[92] H. VERSTEEG AND W. MALALASEKERA, eds., *An Introduction to Computational Fluid Dynamics - The Finite Volume Method*, Pearson Prentice Hall, London, Great Britain, 1995.

[93] B. VOGT, J. STARFLINGER, AND T. SCHULENBERG, *Near Term Application of Supercritical Water Technologies*, no. 89732, Proceedings of ICONE14, Miami, United States, 2006.

[94] WAGNER AND KRUSE (IAPWS-IF97), ed., *Zustandsgroessen von Wasser und Wasserdampf*, Springer Verlag, Berlin, Germany, 1998.

[95] G. S. WAS AND T. R. ALLEN, *Time, Temperature, and Dissolved Oxygen Dependence of Oxidation of Austenitic and Ferritic-Martensitic Alloys in Supercritical Water*, no. 5690, Proceedings of ICAPP'05, Seoul, Korea, 2005.

[96] D. WILCOX, ed., *Turbulence Modelling for CFD*, DCW Industries, La Canada, California, 1998.

[97] J. WRIGHT AND J. PATTERSON, *Status and Application of Supercritical-Water Coolant Reactor*, vol. 28, Proceedings of American Power Conference, 1966, pp. 139–149.

[98] WS ATKINS CONSULTANTS AND MEMBERS OF THE NSC, *Best Practice Guidelines for Marine Applications of Computational Fluid Dynamics*, tech. report, Sirehna, HSVA, FLOWTECH, VTT, 2002.

[99] A. YAMAJI, Y. OKA, J. YANG, J. LIU, Y. ISHIWATARI, AND S. KOSHIZUKA, *Design and Integrity Analyses of the Super LWR Fuel Rod*, no. 556, Proceedings of GLOBAL 05, Tsukuba, Japan, 2005.

[100] A. YAMAJI, T. TANABE, Y. OKA, Y. ISHIWATARI, AND S. KOSHIZUKA, *Evaluation of the Nominal Peak Cladding Surface Temperature of the Super LWR with Subchannel Analysis*, no. 557, Proceedings of GLOBAL 05, Tsukuba, Japan, 2005.

[101] R. ZOBEL, *Experiments for a Hydraulic Elbow*, Mitteilungen des Hydraulischen Instituts, München, Germany, 8 (1930), pp. 1–47.

Appendix

A KTA Guidelines

A.1 Design Fatigue Curves according to KTA 3201.2

A.1.1 Ferritic Steels

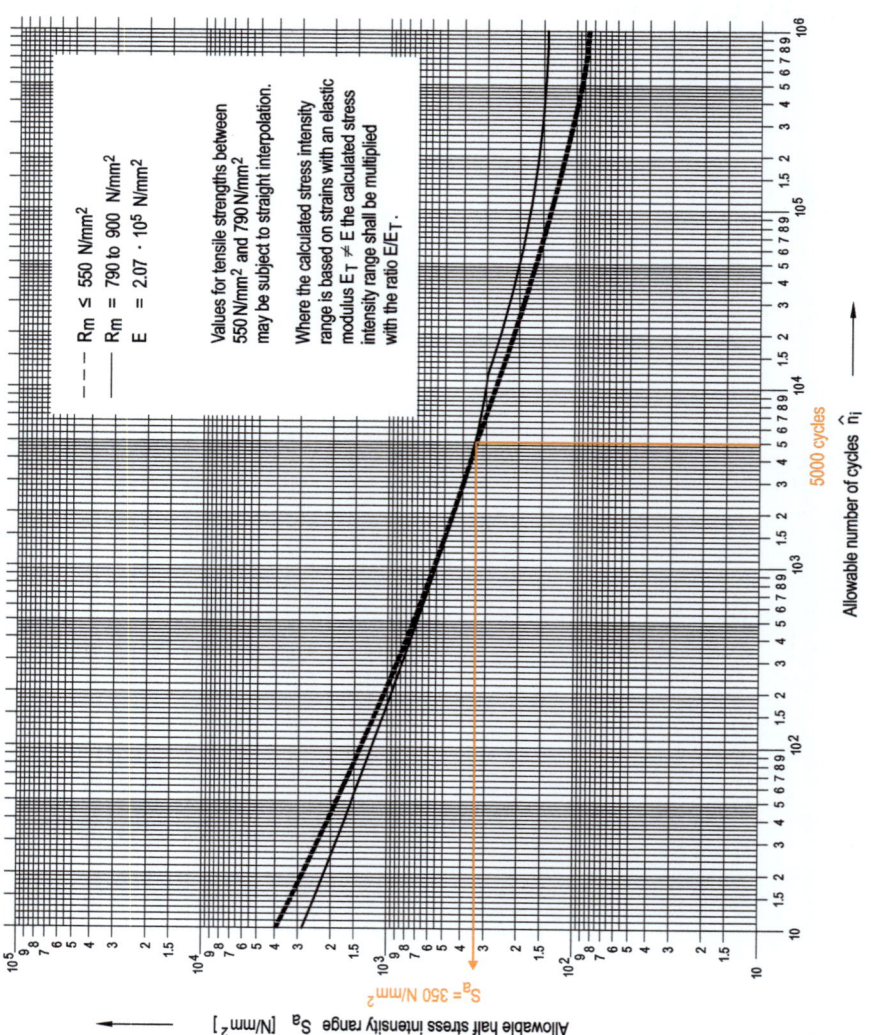

A.1 Design Fatigue Curves according to KTA 3201.2

A.1.2 Austenitic Steels

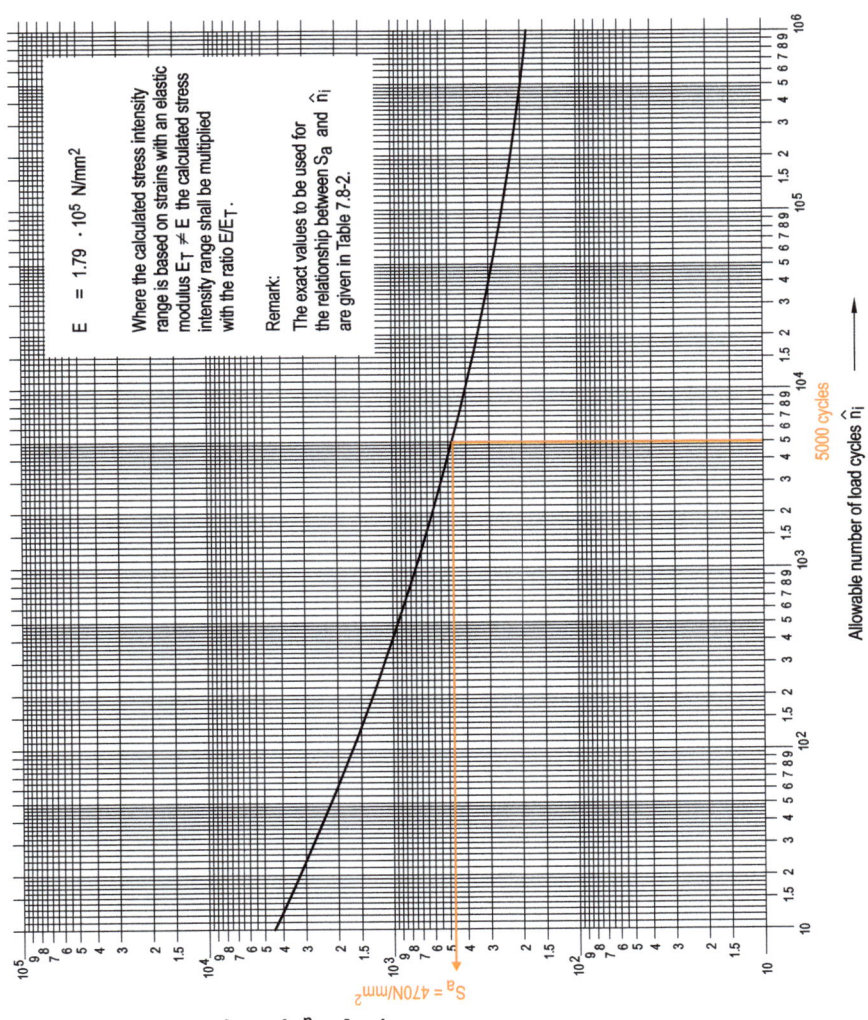

B RPV Assembly for the HPLWR

B.1 Notation

A	Control rod penetration	K	Closure head
B	Dome plate closure head	L	O-ring seal
C	Preloaded spring	M	Backflow limiter
D	Steam outlet	N	Inlet flange
E	Hot pipe	O	CRGA
F	Outlet flange	P	Steam plenum
G1	Fuel assembly cluster (one pass)	Q	Steam plenum support
G2	Fuel assembly cluster (two pass)	R	Vessel support
G3	Fuel assembly cluster (three pass)	S	Core barrel
H	Core support plate	T	Steel reflector
I	Closure head bolt	U	Lower mixing plenum
J	Closure head nut		

B.2 RPV Design for the One Pass Core

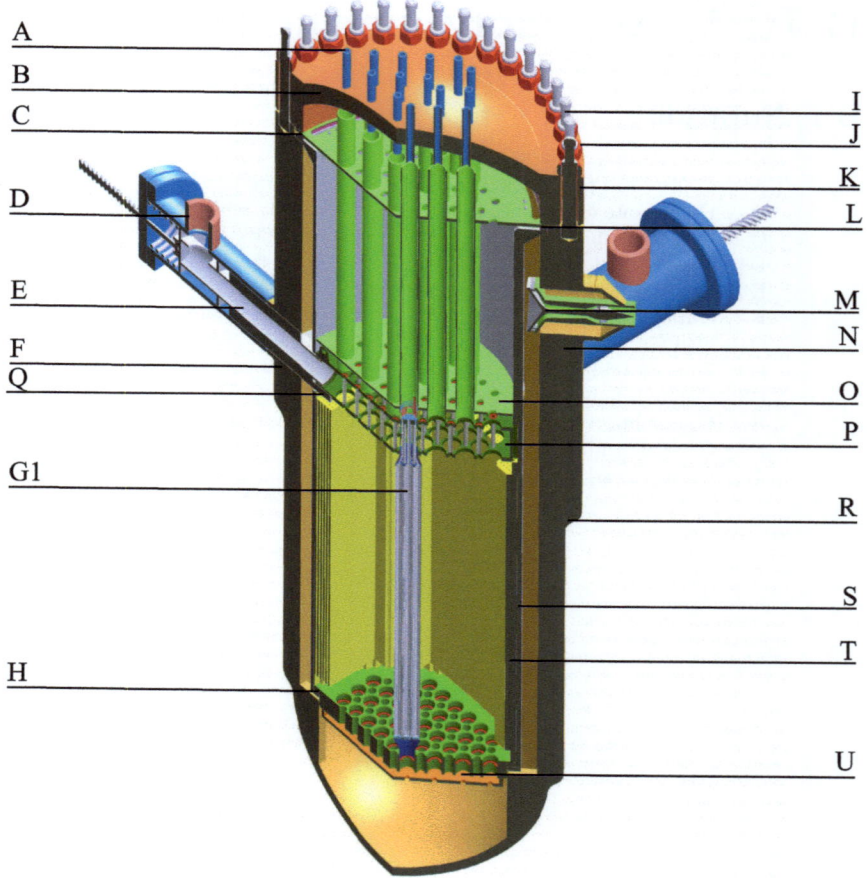

B.3 RPV Design for the Two Pass Core

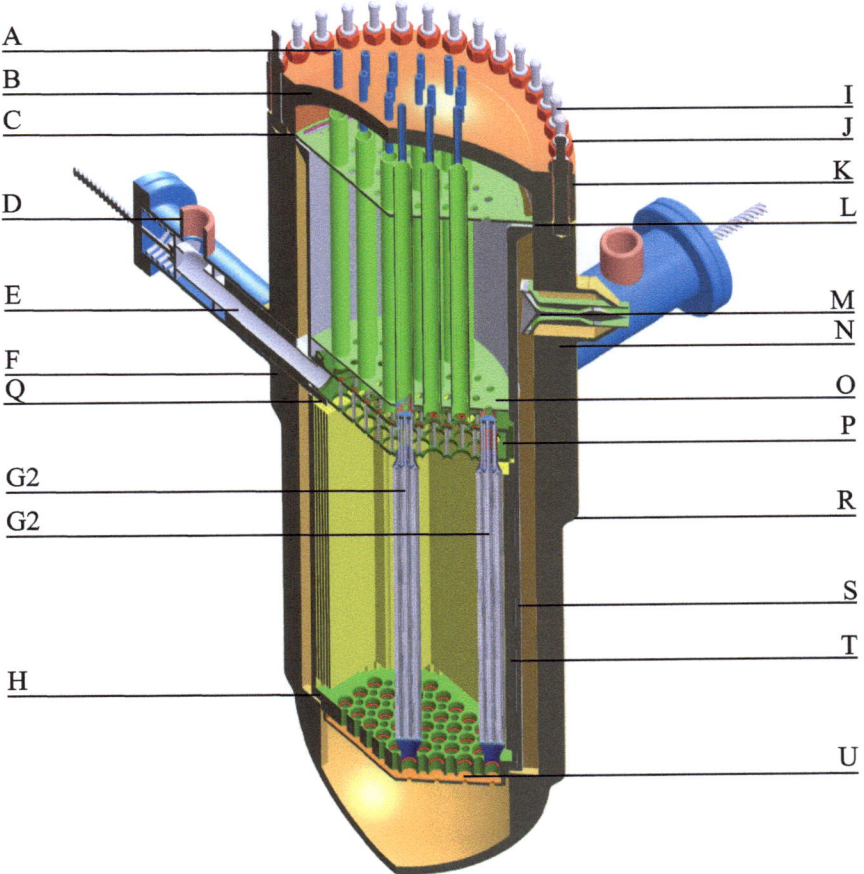

B.4 RPV for the Three Pass Core

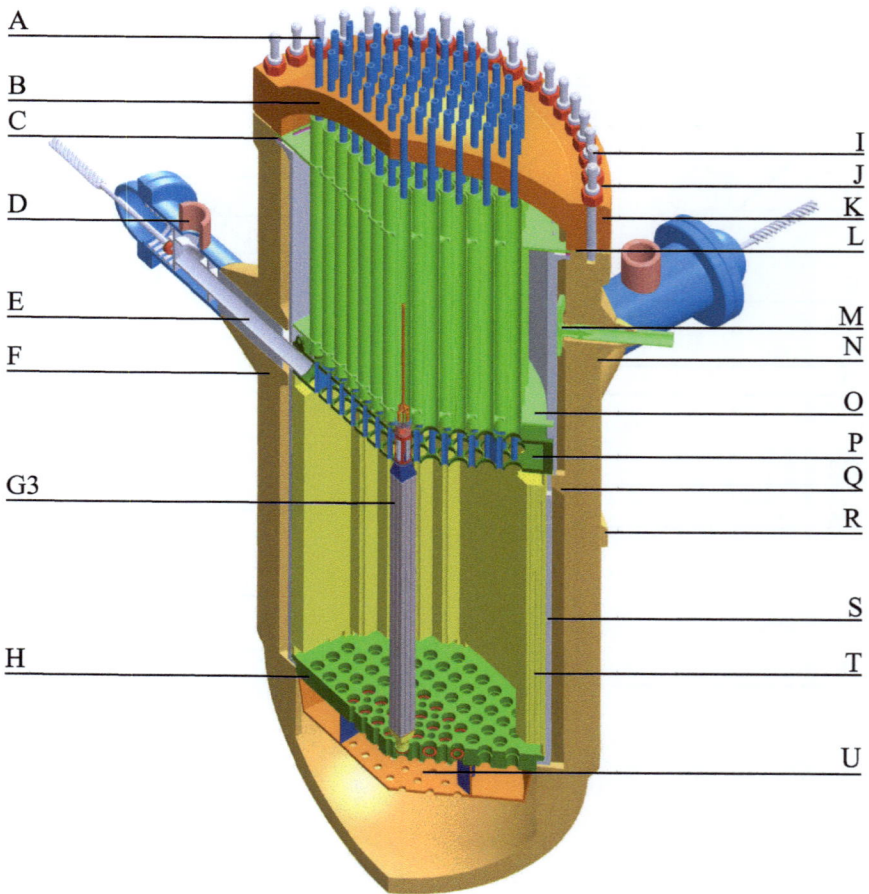

Die VDM Verlagsservicegesellschaft sucht für wissenschaftliche Verlage abgeschlossene und herausragende

Dissertationen, Habilitationen, Diplomarbeiten, Master Theses, Magisterarbeiten usw.

für die kostenlose Publikation als Fachbuch.

Sie verfügen über eine Arbeit, die hohen inhaltlichen und formalen Ansprüchen genügt, und haben Interesse an einer honorarvergüteten Publikation?

Dann senden Sie bitte erste Informationen über sich und Ihre Arbeit per Email an *info@vdm-vsg.de*.

Sie erhalten kurzfristig unser Feedback!

VDM Verlagsservicegesellschaft mbH
Dudweiler Landstr. 99
D - 66123 Saarbrücken

Telefon +49 681 3720 174
Fax   +49 681 3720 1749

www.vdm-vsg.de

Die VDM Verlagsservicegesellschaft mbH vertritt

Printed by Books on Demand GmbH, Norderstedt / Germany